Man's Higher Consciousness

Professor Hilton Hotema

ISBN: 9781639233083

All Rights reserved. No part of this book maybe reproduced without written permission from the publishers, except by a reviewer who may quote brief passages in a review to be printed in a newspaper or magazine.

Printed: September 2021

Cover Art By: Paul Amid

Published and Distributed By:
Lushena Books
607 Country Club Drive, Unit E
Bensenville, IL 60106
www.lushenabks.com

ISBN: 9781639233083

Table Of Contents

Prologue	i
Introduction	v

Lesson No.1

Physical Perfection	01
Breath of Life	01
Stages of Degeneration	02
Magnetism	02
Spiritual Potentiality to Physical Actuality	04
The Way to Improve Physical Man	05
So Called Civilized View	05
Why the Truth is Suppressed	06
Economic Freedom	06

Lesson No.2

The Living Cell	07
Chiropractic Law of Physiology	08
Where Did The Living Cell Come From	08
Is Eating Necessary	09
Elimination	10
Cells Are Not Produced By Food	11
Early Men Were Breatharians	11
The Miraculous Cell	11

Lesson No.3

Physical Immortality	13
Why Man Degenerates	13
The Transportation System	15
Blood Purification	16

Lesson No.4

Vital Adjustment	17
Disease Germs	17
Good Health Is Not Immunity	18
Immunity Reduces Power to Resist	20
Conditions That Destroy Health	21
Danger of Smoking	22
Toleration by the Body	23
Immunity	24

Lesson No.5
- Body Changes — 26
- Misleading the Multitude — 26
- Harmful Practices from Birth — 28
- Fewer Centenarians — 29
- The Body Fights Against Changes — 30
- Rudimentary Organs — 30

Lesson No.6
- Man's Natural Home — 34
- Altitude is Beneficial — 35

Lesson No.7
- She Eats Nothing — 37
- Survival is Nature's Goal — 38
- Breatharianism to Gluttarianism — 39
- Buried Six Months and Lives — 39
- Spent Time in Cell — 40

Lesson No.8
- Materialism — 41
- Weight and Vitality Loss Due to Autointoxication — 41
- Live Without Eating — 43
- The Great Body Normalizer — 43

Lesson No.9
- Body Building Material — 46
- Discovery Amazed Material Science — 47

Lesson No.10
- The Aging Process — 50
- Condition of Artificial Life — 51
- Return Must Be Slow and Gradual — 52
- Antiquity of Man — 52
- How to Reverse Physical Appearance of Aging — 53
- Constipation the National Disorder — 54

Lesson No.11
- Does Man Starve — 56
- Finding the Truth — 57
- Food Stimulates — 57
- It is the Body That Acts — 58
- People Who Crave Poison — 59
- Power of Adaptability — 59
- Fish Does Not Give Brains — 61

Lesson No.12
    Danger of Abrupt Changes — 63
    Men of Great Stature — 63
    Stature Originally Gigantic — 65
    Man's Body Resembles the Planetary Bodies — 66
    Misleading Reports — 67
    Body Craves Food As It Does Poison — 68
    Man Eats to Die — 69

Lesson No.13
    Chronic Auto-Intoxication — 71
    Body Tries to Maintain Balance — 72
    Vitality Increases — 73
    Eat Little And Live Long — 74

Lesson No.14
    Body Needs — 76
    Minerals From Cosmic Rays — 77
    Sensation of Hunger — 78
    Eating Is A Vicious Circle — 78
    Atomic Energy — 80

Lesson No.15
    Eating Poisons — 81
    Dangerous Narcotic From Juice of Poppy — 82
    Mice Unable to Live on Human Diet — 83
    Evolution and Devolution — 85
    Dormant Organs Ready When Needed — 86

Lesson No.16
    Vegetarianism Is Bad — 88
    Body Vitality Reduced — 88
    Most Vegetables Are Not Natural — 89
    Cereals are Bad Foods — 91
    Fruits Easier Produced with Less Labor — 92
    Alimentation and Decrepitude — 93
    Earthy Salts Cause Old Age — 94
    Fruits Have Little Earthy Matter — 95
    Fresh Fruit — 96

Lesson No.17
    Carnivorism Is Bad — 97
    Butter, Milk and Cheese Less Harmful — 98
    Reason For Increased Vitality — 100

Flesh Foods Putrefy	101
Mode of Living Builds Cravings, Aches and Pains	102
48 Million Have Trichinosis	103
Table Salt	104
Opinions on Salt Eating	105
Ancient Wisdom	107

Lesson No.18

Longevity	108
He Lived 256 Years	115
Body Never More Than Seven Years Old	116
Fruit and Longevity	117
Doctors Do Not Live Long	118
Live 200 to 300 Years	119

Lesson No.19

Water Causes Aging	121
Less Minerals Needed After Maturity	122
Causes of Sclerosis	122
Importance of Water	123
"Mineral Water" Not Beneficial	124
Lime Deposits Cause Stiff Joints	125
Rain Water and Distilled Water Are Safe	126

Lesson No.20

The Wonderful Orange by Dr. Leon A. Wilcox	128
King of Fruits	129
Finest Distilled Water	129
Six Months on Orange Juice by John W. Marshall	130

Lesson No.21

Breath of Life	136
Early Theories of Respiration	136
Physico-Chemical Theory of Respiration Rediscovered	137
Breathing Primary Function	139
Why Man Dies	140
Why Man Lives	140

Lesson No.22

Spiritual Organs	143
Materialism Is A Superstition	144
Recovery From Illness Only Partial	145
The Kingdom of God Within	146
Uncanny Powers of Indians	147

Lesson No.23
- Spiritual Powers — 150
- Ancient Science of Man — 151
- Spiritual Intelligence — 151
- Man's Intelligence — 153
- Intelligence of Animals — 153
- Man A Miniature Universe — 155
- Man Is Dead As He Lives — 155
- Man Lives in the Spiritual World — 156
- Parthenogenesis (Virgin Birth) — 157
- The Kingdom of God — 158
- Feeble Minds — 160
- Ancient Science vs. Modern Nonsense — 161

Lesson No.24
- Physical Purification — 163
- Function of Breathing — 164
- Shower of Red Mist — 165
- Blood Poison — 166
- Breath Culture — 167
- Vital Function — 168
- The Residual Air — 169
- Gases and Acids — 169
- The Skin — 171
- Exhalation — 171
- Atmospheric Pressure — 174
- 500 Felled by Gas — 174
- Suffocation — 175
- Cold Facts — 175

Lesson No.25
- Breath of Death — 177
- Poisons in the Blood — 178
- Deadly Carbon Dioxide — 179
- Exhaled Air Is Poisonous — 180
- Poisons Entering the Body — 181
- Big Battle for Health — 182
- Millions Chronically Ill — 183

Lesson No.26
- Poisoned Air — 184
- Carbon Monoxide Gas — 185

Dangers of Poisoned Air Unknown	186
Brain Poison	187
Pernicious Anemia	188
Cerebral Hemorrhage	189
Lung Cancer	190
Nerve Cells Destroyed	191
Black Lungs	191
Causes Cancer	194
Eat Up the Body	194
Smoke, Soot, Acid, Gas	195
Tobacco Smoke	197
Coronary Thrombosis	198
Science Editor Dies of "Heart Attack"	199
Cigarette Consumption	200
King George VI	200
Change Your World	201

Lesson No.27

The Common Cold	203
Degenerative Process	203
Polluted Air	205
Acute Ailments	207
Air Cells Burst	208
Vital Adjustment	209
No Complete Recovery	210
Hardened Mucus	211
An Invisible Foe	212
The Aging Process	213
The Blood	213
Wonders of the Air	215

Lesson No.28

Cosmic Air Purifies	217
The Home	219
Where to Live and Sleep	219
Ozone	220
Ionized Air	222
Shallow and Deep Breathing	223
Breathe More — Eat Less	225
Air is Life	226
Eternal Physical Life	227

The New Age	227
Lesson No.29	
Degeneracy of Civilized Man	232
Civilization	234
Alexis Carrel	235
Lesson No.30	
Mysteries of Life	238
Modern Science Is Young	239
Life	240
Science Changes With the Wind	241
Atheists and Evolutionists	242
The Physical	242
The Cosmic Circle	243
The Physical Plane	245
Four Cosmic Bodies	246
Rise Above the Material	247
Physics and Chemistry	247
Intelligence	248
Fourth Dimension	249
So-Called Law of Gravitation	251
End of Creation Theory	252
Nature	253
Two Kinds of Facts	255
See the Invisible From Within	256
Science of Man	257
Ancient Wisdom Destroyed	257
Ancient Science	259
Spiritual Kingdom Within	260
The Ancient Voice	261
The Glorious Resurrection	262
Lesson No.31	
The Sacred Science	263
Dangerous Invention	263
More Pious Fraud	264
Spiritual Life	264
Regeneration	265
Man's Decreasing Powers	266
The Red Dragon	267
Two Laws of Generation	268

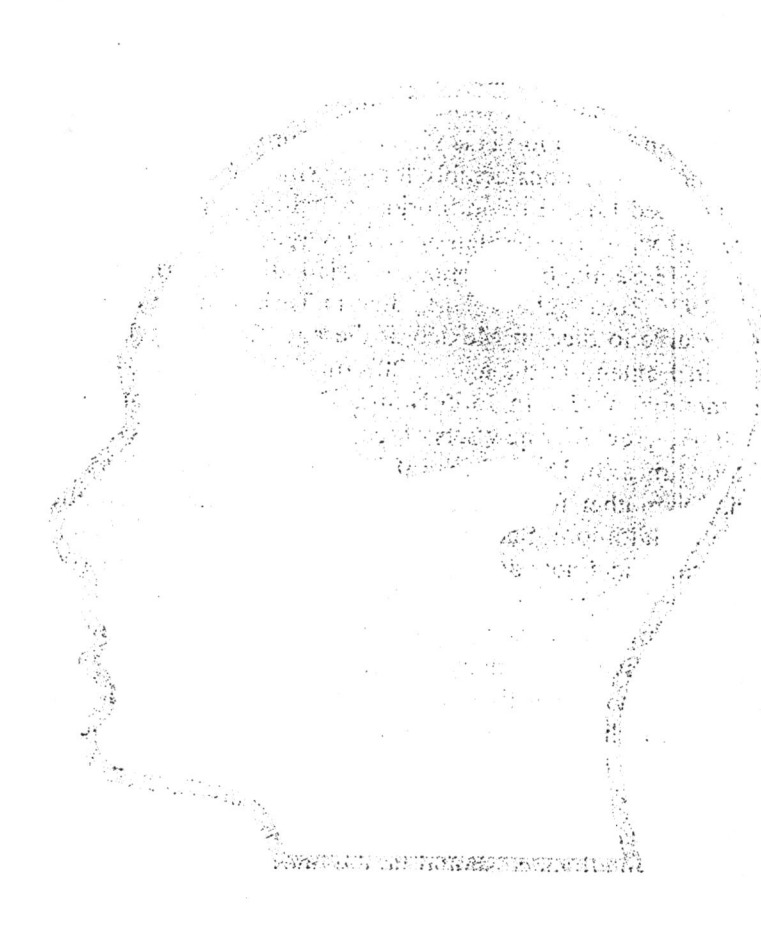

Generation and Death 268
Seven Spiritual Centers 270

Prologue

Life is Creation's greatest treasure for Man in the flesh, and most men should enjoy it much longer than they do. This can easily be done by learning the body's simple requirements, and living in harmony with that knowledge, which is contained in Hotema's late work titled Long Life in Florida. Some men are now living 120, 150 and 200 years and even longer, and what is possible for one man is possible for millions more. Charlie Smith of Florida is vital and vigorous at the age of 119 and says he intends to live considerably longer. Much about him is contained in our work titled Long Life in Florida, which everyone should read, as it's the greatest work ever written on Longevity.

In 1943 Santiago Serviette, an Indian, died in Arizona at the age of 135. In 1936 Zora Agha, a Turk, died in Turkey at the age of 162. In 1921 Jose Calverio died in Mexico at the age of 185. In 1795 Thomas Carn died in England at the age of 207. In 1933 Li Chung-Tun died in China at the age of 256. In 1566 Numes De Cugna died in India at the age of 370. He grew four new sets of teeth and his hair turned from black to gray four times. In December, 1888, when Mrs. Fred Miller was two years old, her mother found a little turtle nearly frozen in an alley near their home in Baltimore. She took the turtle home, named it Pete, nursed it back to health, and now at an age estimated to be more than 100 years, Pete is still the family pet and shows little signs of his age.

All living creatures are ruled by the same laws of Creation, but they do not all live in harmony with those laws. Those that reach the closest to it are those that live the longest in proportion to the length of time required for them to reach maturity of physical development. There are many reasons, most of them preventable, why people die young and hospitals are filled with the sick, whereas others are seldom ill and live three to ten times longer. It would logically seem that living creatures with the higher intelligence should be the ones to live the nearest to the requirements of the laws of Creation; but in action it seems to work the other way, the more intelligent creatures being those who appear to stray the farthest from the straight and narrow path which leads unto life (Mat. 7:13, 14). The large majority of the so-called health writers are not noted for longevity. They seem to die as early as those who read their writings. And some who live 120 years can't tell why they lived so long, as in the case of Diamond, who lived 120 years, wrote a book on long life in 1904 when he was 108, on page 43 of which he itemized "My Daily Menu", which is rather good, yet not one we would recommend.

To reduce longevity to a scientific basis means that we must learn the requirements of the body and supply them. The facts show that living is breathing. We can't die as long as we can breathe — we stop living when we stop breathing. Breathing is the primary function of the body. We can live for weeks without eating, and for days without drinking, but when we stop breathing for a few minutes we stop living.

And right here is the most neglected spot in the entire health field. Why is that so? First, ignorance: second, the claim of science that man lives on what he eats; and third, no one has yet found a way to commercialize air end breathing. This is the discouraging condition we found sixty years ago, when we set out to learn how to live healthy and long. And so, we began almost alone to learn something about breathing and the Breath of Life. The first valuable hint came when we found this: "If we maintain our blood in normal condition and circulation, sickness would be almost impossible. The blood is the life of the flesh. We are what we are by the influence of our blood flowing through our body" (Bernarr Macfadden, in Vitality Supreme, 1910).

Then there came another surprise when we discovered that blood is made of gas. The gases of the air constitute the total composition of the blood. We know that water is the product of the uniting of hydrogen and oxygen gases. When we drink water, we drink gases in fluid form. Blood is gases in fluid form. Everything can be transformed to gas by heat. The earth itself is constituted of condensed gas. This is the source and origin of all things. We have heard of fire damp, ignis fatuus, and will-o'-the-wisp. That fiery element is the Living Gas in what we eat and drink. That Living Gas is all the body uses of what we eat. Of that Living Gas the blood is made. This is the first valuable lesson in dietetics. Scientists talk learnedly and foolishly about protein, carbohydrates, nucleic acids, fats, lipids, etc., ignoring the fact that the ox, elephant, horse and moose live in good health all their days on grass and green leaves.

The next lessons in dietetics is not to heat food and drive out of it the precious, volatile gases which the body uses in its laboratory to make its blood and the products it needs, which includes all the elements mentioned above. Remember this one: Creation never uses second-hand material in its building work. The protein in the food we eat never becomes the protein of our body. That protein has served its purpose, is a used product, and is never used again by Creation in its constructive work.

The living gas in what we eat is all the body uses. The rest is useless waste, and cast off by the body as feces. Hence, most of what man eats goes down the sewer. As gases are all the body uses in making blood and bone and building flesh, consider the condition of the blood, bone and flesh that are made of the poisonous gases that saturate the air of Modern civilization, where health is the exception instead of the rule. If a chemist analyzed the air we breathe and gave us his report, we would be shocked to learn the great amount of poison the body must endure to live in our polluted environment. This subject is so broad and vital, it would take a large book to discuss it adequately. But enough has been said to make a thinking person be more careful about the kind of air he breathes, the condition of the air in his home, and especially in his bedroom, where lack of activity during the night allows the air to stagnate and grow extra foul. That's another reason why people die in their sleep. The polluted, stagnant air in their bedroom paralyzes the breathing center of the brain, and they just stop breathing.

Stagnant air gets foul, like water in a stagnant pool. Keep the air in circulation in homes and bedrooms. Use electric fans for that purpose. Fewer people would die in their sleep if they had an electric fan in operation in their bedroom. This writer is only fifteen years under the century mark, which he expects to pass by many years, as he feels as fit as he did forty years ago. And he is telling the world here, in this work of his, some of the secrets of how he has done it.

To an intelligent, unprejudiced person who can and does think, the information contained in this work may seem simple. But it is the fundamental simplicities that are always difficult to accept, because they are so very simple, and, therefore, unbelievable on that account.

Prof. Hilton Hotema
Tequisquiapan, Qro.,
Mexico.
Alt. 6000.

Why We Live & Why We Die

Scientific investigation shows that —

We breathe to live and we breathe to die,
We drink to live and we drink to die,
We eat to live and we eat to die.

In this work it is sought to show how to take DEATH out of Breathing, Drinking and Eating, so man may live 250 to 300 years. As Dr. Robert McCarrison, of the British Army Medical Service in India, reported of a colony of people he found living in a certain Himalayan region, who were active and vigorous, yet so old in years that he was astounded and could not believe their records were correct. But he found no error in their way of keeping time. He said: *"Men well over 200 years of age working in the fields with much younger men, doing as much work, and looking so much like the younger men, that I was unable to distinguish the old from the young."* - K. L. Coe in Correct Eating &Strength, March, 1931.

Introduction

The living cell is animatized by Cosmic Force and Intelligentized by Cosmic Consciousness. The human body is a mass of trillions of cells, guided by Infinite Intelligence. Each cell is a mass of millions of atoms, each of which is a miniature solar system, with "planets" in the form of tiny electrons whirling at tremendous speed around a common center of attraction. The solar system has no use for food. The electrons do not eat. The atoms do not eat. The cells do not eat. Why should man eat?

The Parent Cell begins the body of man and expresses the uncanny intelligence which it inherits from the Cosmic Principle of Creation. This cell does not come from food, and does not depend on food. The body is not the product of food, and should not depend on food. Food cannot sustain what it cannot produce. As the body is not the product of food, it is actually not sustained by food. Infinite Intelligence works in the cell, from the Parent Cell on through the entire existence of the body. Being directed by infinite Intelligence, the cells know spontaneously the functions they are to perform in the construction and maintenance of the body. The innate intelligence of the part the Cell must play in the whole body is a mode of being of all the elements of the body. These elements know their work, and can receive no aid from human hands.

The body cells are seen to understand numerology, geometry, physiology and biology, and act concertedly for the good of the whole. The spontaneous tendency of the cells toward the formation of the organ in the body is a primary datum of observation.

The body, from the Parent Cell on, is built by techniques and directed by Intelligence entirely foreign to the best scientific minds. From the Parent Cell composed of invisible atoms, and out of the invisible Essence of the Universe in the Infinite Air which contains ALL in itself, the body and its organs come into being by the work a cells endowed with Infinite Intelligence. The cells prove by their work that they possess a prevision of the future structure and its purpose, and they synthesize from the atomic substance that appears to be contained in the plasma, not only the building material, but also the builders.

The trillions of cells forming the body are tiny suns and stars, composed of the same cosmic essence and governed by the same cosmic law. A droplet of water forms a tiny microcosm containing a great variety of cosmic chemical elements. The body cell is practically a duplicate of a water droplet, but raised to the high plane of divine animation.

If the cells depend on food and drink, man should continue to grow and live as long as he had sufficient food and drink. According to the press of May 27th, 1937, Srimati Bala, of Bankura, India, age 88 had taken neither food nor water since she was 12. The account said "She is always gay and looks like a child." What is possible in one case is possible in millions of other cases. Abbe N. De Montfaucon De Villars stated in his book that the Ancient Masters ate food only for pleasure and never of necessity (Comte De Gabalis).

Man Eats To Die

The world is flooded with books on food and feeding. No one seems to realize that eating is not natural, but an acquired habit, like smoking and drinking, and that Air is the Cosmic Reservoir of all things, including the substance that builds and sustains the human body. Science shows that the body is built of cells, which are composed of molecules, which are composed of atoms, which are composed of electrons, which are nothing more than whirling centers of force in the ether.

Electrons do not eat, atoms do not eat, molecules do not eat, cells do not eat, and the body is built of and sustained by the cells, and not what man eats. More proof that eating is not only a habit but a bad one, appears in the fact that a sick man often begins at once to recover when given no food, and even shows signs of growing younger.

This could not be, and it would be dangerous for one to fast, if eating was natural and food was needed to sustain the body. Why does man seem to starve to death when deprived of food? That riddle "MANS MIRACULOUS UNUSED POWERS" considers. For more than half a century, the author read books on food and feeding, and closely followed the arguments and explanations. He found those who favored Vegetarianism omitted all the bad features, and the same course was pursued by those who favored Carnivorism. Books favoring Vegetarianism say nothing of the damaging qualities of vegetables and Cereals. Those favoring Carnivorism carefully omit the damaging properties of flesh. These authors lead their readers astray with half truths. A half truth is more dangerous than a lie, as it is more misleading.

The author describes the damaging qualities of all foods, and favors none. What he shows will shock the reader and show him why that group of eminent doctors in the 19th century, after studying the food question from every angle, concluded with this astounding statement:

"We eat to live, and we eat to die."

Why is it possible that we eat to live and eat to die? If we eat to live, how can we eat to die? If we eat to die, how can we eat to live? In this work these puzzling questions are considered and answered.

Lesson No. 1
Physical Perfection

Physical Perfection would be Physical Immortality; and Herbert Spencer formulated the Law of Physical Immortality as follows to wit: "Perfect correspondence would be perfect life. Were there no changes in the Environment but such as the organism had adapted changes to meet and were it never to fail in the efficiency with which it met them, there would be Eternal Existence and Eternal Knowledge" (First Principles).

Breatharianism in Physical Perfection: Man came into physical existence a Perfect Breatharian. God breathed into his nostrils the Breath of Life, and man became a living entity (Gen. 2:7). Nothing was lacking, and nothing more was needed. The Breath of Life supplied all the requirements of animation. The Breatharian needs air only, and nothing more, to sustain his body. In that state of Physical Perfection man has no other wants. The less man needs the more he becomes like gods, who use nothing and are immortal, said the Ancient Masters.

Poverty, Want and Sickness are the work of man. They are the products of his habits which correspond with his desires. He increases his burdens as he multiplies his wants. The less man needs the more complete he becomes. He gains Perfection as he gains freedom from all Wants. The more Wants he has, the less complete he is, and the farther he inclines from Perfection. He changes his world as he changes himself.

Breath of Life

"Every living thing must breathe the air in order to live. The tree breathes the air through its leaves. The leaves are in this sense the lungs of the tree. Insects breathe the air through tiny openings in their bodies. Frogs breathe the air partly through the skin. Fishes breathe the air by taking oxygen out of the water as it passes over their gills. Man breathes the air through the air cells of the lungs." Frederick M. Rossiter, B.S. M.D., L.C.R.P., London. "Hardly anyone understands this science (of breathing), which all should know and practice. It is the air that renews our blood and brings life to all our organs. It is the air that helps to give us balance and to keep our physical and psychic functions in good order . . . Most people use only a third, a quarter or even a fifth part of the lung's total surface." - Professor Edmond Szekely.

Prof. Hotema   *Man's Higher Consciousness*   *Lesson No.1*

Modern Man is the degenerate descendant of the Breatharian. During the millions of years that man has inhabited the earth, his environment and the many habits he has formed have forced his body to adapt itself, under the law of Vital Adjustment, to many evil conditions and harmful substances in order to survive, all of which are foreign to the body and injurious in character.

Stages of Degeneration

Every facet of living existence, the Law of Vital Adjustment, the evidence contained in ancient scriptures, all prove without an exception that Modern man is the product of descending Evolution — Devolution. The scanty evidence that has survived from the remote past, shows that Modern man has slowly descended through the five stages:
1. Breatharianism 2. Liquidarianism 3. Fruitarianism 4. Vegetarianism 5. Carnivorism.

Breatharian indicates a plant or animal that neither eats nor drinks and subsists entirely on the substances contained in the air. It is surprising to know how many such plants and animals there are. The Spanish moss hanging on the trees of Florida gets all its substance from the air, and grows very fast. The big cactus plants on the dry, barren deserts of the southwestern part of the U.S.A. get their substance from the air, their roots serving as anchors to hold the plants in place and complete the Life Circuit.

Magnetism

In our work, "THE NUTRITIONAL MYTH," we stated that no chemist can find in the grand in which grows a giant tree, the ash, mineral, carbon, wood and chlorophyll contained in the body of the tree and its leaves. Nor does a tree consume the soil in which it stands. If so, as evidence of such consumption a depression should surround every tree. A large tree weighs tons, and all the earth still surrounds it that was there when it was a small sprout. The roots appear to serve only as anchors to hold the tree in place and complete the electro-magnetic Circuit of Life.

If corn or wheat is planted in moist sawdust around a magnet or loadstone, growth is promoted without adding more elements. No real relation appears to exist between the elements in the soil and the plant. Examination shows that plants contain elements which never exist in the soil. They are supplied by cosmic rays.

It is the general belief that crops consume the soil in which they grow. Jean Van Helmont did not believe it. In the 17th century he weighed the soil he put in a tub, then planted a tree in it. At the end of four years the tree was six feet high and weighed many pounds. But the soil in the tub weighed the same. The growth of the tree resulted from the elements supplied by cosmic rays. There must be ground connections in the base of plants to complete the Life Circuit. Ground connections are necessary to complete the circuit of electric instruments.

Trees grow not in certain areas, nor grow well in what is termed poor soil. They improve if given proper fertilizer. To make the circuit effective, the right minerals must be in the ground, and there must also be moisture. It is not the tree that needs the minerals. They are required to supply the conditions necessary for proper electro-magnetic action. You know more about motor car batteries. When the battery weakens and runs down, it is recharged. The recharging process does not infuse electricity into the battery. It only changes the chemistry of the battery fluid.

All hibernating animals are complete breatharians during the weeks they sleep in winter. A fasting man who takes only water during his fast has simply added liquidarianism to breatharianism. That is the path back to Physical Perfection. In our work "THE NUTRITIONAL MYTH" we referred to cases of those who are reported to live without eating — Mrs. Martha Nasch who had eaten nothing for seven years; Teresa Neumann, a German mystic, who had taken no food nor liquid for forty years.

The press of May 3rd, 1936, mentioned the cases of Srimati Giri Bala Devi of Patrasayar in Bankura, India, the sister of Babu Lambadar Dey, pleader, as a woman who had taken neither food nor water for 58 years. The account said: "She takes nothing, not even a drop of water. She is always gay and looks like a child. She does not pass stool or urine, and does her house work like any other woman." This woman was reported as being 68 years old, yet she "looks like is child." This appears to be getting close to the Law of Perpetual Youth and Eternal Physical Existence.

One physiologist states that of what man eats, the only parts that do and can enter his blood are the gaseous and fluidous elements. All the rest is nothing but waste that passes through the bowels as feces, playing its part in overworking and weakening the stomach and bowels, with constipation afflicting practically the whole nation. There are very few folks who are free of stomach and bowel troubles, and the markets are flooded with worthless and harmful remedies for these disorders.

Spiritual Potentiality to Physical Actuality

Scholars hold that in Physical Perfection man was entirely free of all wants and desires. The needs and requirements of his body were fully supplied by the Pure Air of His Perfect Environment.

That perfect state was a condition precedent to his coming into physical being. It was the cosmic perfection of his Environment that made possible his evolution from Spiritual Potentiality to Physical Actuality.

It is an established fact that something cannot come from nothing. Men could not become a physical entity had he not existed as a spiritual potentiality. There must be a precedent for every subsequent, a cause for every effect. The cause, whether first, last or anywhere along the chain of causes, must be the comprehensive equal of the effect.

The stream cannot rise above its source. If the rivulet can flow but an inch higher than the sufficiency of its cause, there is no reason why it should not climb the mountain-top and "increase by the force of its own intensity," as is said of disease. No effect can produce its own cause. The Universe could not create itself, which is equivalent to the admission that no part thereof could create itself.

Universal existence is eternal existence. What appears as the physical world which we call Nature, is the materialization of Spiritual Potentialities. Had man not existed Spiritually, he had never become a physical reality; and his appearance as a physical entity is conclusive proof of the perfection of his Environment at the time of his Physical beginning. That Environment had to and did possess the perfect powers and requirements that evolved man, not from an ape, but from Spiritual Potentiality to Physical Actuality. It fulfilled every need, every demand, everything, it had been fatal.

The Way To Improve Physical Man

Advanced scholars point out that there could have been in man's physical beginning no unfilled wants; otherwise physical being had been impossible. They show that the only way to improve physical man, is to reduce his wants and decrease his economic burden.

But physical science is not interested in any course that raises man to a higher plane. For nothing must be done, nothing must be allowed to happen, that will disturb or derange the fixed social order of civilization.

This order is the product of ages of planning and scheming. It depends upon and is sustained by man's wants and desires, and the constant effort made to promote and increase them. To that end all education is directed. Every branch and department of civilization leads away from Perfection. The movement away from Perfection begins with the child in school and continues all through life.

So Called Civilized View

On January 12th, 1951, Frank W. Abrams, Chairman of the Board, Standard Oil Company of New Jersey, made an address before the National Citizens Commission for Public Schools, and his address was published and widely circulated. Among other things he said: "There can be no doubt that we are talking about something very fundamental to business when we talk about education.... If only to maintain and expand its markets, the business world has at least as big a stake as anyone in the achievement of an educated, productive and tolerant society.... There is definite correlation between education and the consumption of commodities. Education has done more to create markets for business than any other force in America."

This is the orthodox view, and according to that view, the purpose of education is to maintain and expand the markets of business, and to create a demand for commodities. To that end billions of dollars are expended annually in the education of the children of America.

The constant cry of Commercialism is to consume more, create new markets and new demands, promote the production of commodities, employ more wage slaves, and increase the economic burden.

Why The Truth Is Suppressed

The art of scientific living is so lightly regarded that it receives no attention. He who is so far ahead of the multitude as to oppose the social pattern is promptly silenced, disgraced and liquidated; and the press carries large headlines proclaiming that an enemy of social progress has been found and jailed. The deceived multitude believes.

One's teaching may be in harmony with God's Plan of Life, the Law of Perfection, and the Science of Cosmic Economy. But that teaching does not harmonize with civilization's artificial world, or support its social pattern; therefore, it cannot be accepted and supported by any institution or any form of government. It must be suppressed *"for the good of the people."* Tell us how long man's created wants and unnatural desires will mean money for Commercialism, and we will tell you how long man will remain in his present condition of degeneracy and economic slavery.

Economic Freedom

It is interesting and important to note that as man moves back toward primal Perfection, his wants decline and his economic burdens decrease. We thus learn what these burdens are and whence they came. We see them as the product of man's created wants and unnatural desires which Perfect Man had not. Man has produced them, and he can destroy them. It was not until man began to form habits and adopt practices which created wants that he began to decline and degenerate. He was deceived then, as he is now, by the illusion of progress as he developed new habits that increased his wants. He considered then, as he does now, that each new invention was a mark of progress while he and the doctors were puzzled by the fact that his health continued to decline and his life-span to decrease. Economic freedom is the first step back toward man's high estate of primal Perfection. Every animal, in its native state, has economic freedom. Man is the only economic slave on earth. He has made himself such by his unnatural wants and acquired desires.

In complete freedom from every want, to be dependent upon nothing, man's mind and senses are under control. He is released from the consequences of action, which are bonds and chains, binding down those who are the slaves of want and desire.

Lesson No. 2
The Living Cell

"Life is the expression of a series of chemical changes." - Osler in Modern Medicine P. 39. Modern science has spent years trying to define Life, and some of the many definitions advanced appear under "Life" in our work titled "SCIENTIFIC LIVING."

In "The Book of Popular Science" it is said: "What is Life? That is another question and the answer is one of the profound mysteries. Science has explored life down to a single cell of living matter, but exactly what makes the cell alive is not known."

One of the main concepts of science, the "living protoplasm" which has been regarded by science for a century as the source of life, was exploded in 1937 by group of America's foremost scientists, who showed that protoplasm is composed of numerous, ordinary particles, visible by modern methods, and not one of these particles are alive. Protoplasm is the sticky, whitish substance of which living tissues seem to be made. It was so named in 1840 from the Greek Protos (first), and plasm, a thing formed, and means "the first citation."

Dr. E. N. Harvey, of Princeton University, related how, under the microscope, protoplasm has literally fallen apart in the whirling centrifuge and fine testing methods. It appeared to be composed of cells, bits of fat, granules of colored matter, proteins, threads, hollow bubbles, nuclei, minerals and a complex of other substances, — "not one of which," he said, "can be considered as living except insofar as it is indispensable for the continuance of Life." Thus, the mystery of Life remains unsolved so far as Modern science is concerned. Osler declared that Life is neither an entity nor a principle, but only the expression of a series of chemical changes. Then came the famous Carrel whose experiments showed that Osler is wrong. Carrel said, "The childish physico-chemical conceptions of the human being, in which so many physiologists and physicians believe, have to be definitely abandoned" (Man, The Unknown, P. 108).

This leaves Medical Art with no law of Physiology, and it cannot have one until it knows what makes the body function.

Chiropractic Law Of Physiology

The Science of Chiropractic discovered the law of Physiology, and the law was formulated by the great Willard Carver, who died in 1943 at the age of 80. He founded the Carver Chiropractic College at Oklahoma City in 1905. Carver wrote: "Organisms that move in conduct we term animation, are not alive. Life is imminent in such structures, but not inherent in them. Life flows through them. We speak of a wire as being alive whilst a current of electricity is flowing through it. If we cut off the current, the wire becomes dead; yet the wire itself has assumed no visible change. "Life, as it is observed in this material existence, is the action of Matter under the operation of Forces. The body of man is alive in the sense that it is animated by the Life Principle flowing through it." - (Psycho-Bio-Physiology).

According to Carver, the Living Organism is an instrument and its animating power is the Life Principle. But this old doctrine of the Ancient Master, who held that the Cosmic Spirit animates the body of man (Jn.6:63), "is a mere superstition," according to Modern science, and accepted now only by the layman and the Chiropractors. No satisfactory theory of the evolution of man, as advanced by the evolutionist, can be sustained so long as the Genesis of Life upon this planet is shrouded in darkness. The basic factors and causes of evolution are bound up in the question of Life itself.

The evolutionist begins with the Living Cell. Once given a primordial life all cells are capable of reproduction, and modern science constructs man physically, vitally, intellectually and morally. According to Modern science, all these properties appear in matter as the result of "the expression of a series of chemical changes." Preposterous: that exceeds the work of the magician in pulling white rabbits out of a hat.

Where Did The Living Cell Come From

The evolutionist makes no attempt to explain the original appearance of the living cell itself. Neither does he explain the nature nor cause that originated it. He does not explain the original division of living organisms into male and female. He does not explain the phenomenon of intelligence displayed in the conduct of all living things.

Modern science explains none of these things. On the other hand, it most unscientifically relegates them to the realm of the "Unknowable."

Modern science confesses itself baffled at every point when it would explain how life evolves from non-life, how sensation evolves from non-sensation, how intelligence appears in living creatures. It fails to explain these phenomena just as it fails to explain how intuitive intelligence rises into rational intelligence, or how unmoral perceptions rise into moral conceptions. The vigilant biologist traces life to the nucleated cell. Here, in the department of Protozoa, he becomes bewildered. He misses the connecting link. He fails to discover that subtle principle which enters into and converts inanimate substance into living organism. This lack of basic knowledge caused Carrel to state that, "Our ignorance (of the body and its functions) is profound.... The mechanistic physiologists of the 19th century... have committed an error in endeavoring to reduce man entirely to physical chemistry (Man, The Unknown, P. 4, 34). In more definite terms, modern physiologists of the orthodox school are still in darkness as to the body's requirements and functions. What the body needs to sustain it and why it functions are things unknown to them. The field is wide open for the consideration of Man, with almost nothing definitely known by physical science, and any reasonable theory advanced may be as correct as any other.

Is Eating Necessary

"Appetite comes with eating." - F. A. Ridley.

In our work titled "THE NUTRITIONAL MYTH," we stated that it is more difficult to explain why man should eat than to show that he should never eat. It is said by some unorthodox physiologists that man should not eat; that what he does eat does not sustain his body; that the cells of which his body is composed are self-existent and self-sustaining, as are all other objects that are composed of atoms as the body cells are; and that man should have no more need for food than a stone or a star, since these also are constituted of atoms the same as are the body cells.

Science knows that the Parent Cell begins the body and builds it by a process of cell division and subdivision. Food does not build the cell or the body. Neither can food sustain the cell nor the body.

New cells are not produced by food but by the division of pre-existent cells. The new cells replace the disintegrated cells; modern science knows that. It knows that the new cells are not the product of food. Yet it insists that the body is nourished and sustained by what man eats. Consistency of thought demands that we proceed in our processes in a direct line through infinite time to infinite results. If food does not produce cells, if all cells of the body are produced by the process of division and sub-division of pre-existent cells, then food does not and cannot nourish and sustain the cells. In that case the cells need no nourishment, cannot receive or use it, and are self-existent and eternal.

If food can produce and sustain Living Cells, if unintelligent matter can produce intelligence, there is no reason why man should not progress to infinite capacity by virtue of the power residing in the food he eats.

Elimination

Experiments consistently show without an exception, that elimination is much more important than feeding. Many refuse to recognize that fact. The maintenance of the vital condition of the body is more intimately and immediately related to and dependent upon the excreting part of the physiological process than it is upon the supply of new aliment. Many refuse to recognize the fact that feeding may be suspended for a considerable period without causing anything more serious than loss of weight and strength. The records show that people have lived without food all the way from forty days to forty years, and still lived in good health. But the elimination of the effete substance produced by tissue disintegration cannot be checked for even a few minutes, in warm blooded animals, without inducing fatal results.

Every act of respiration is in effect the leading process of excretion, and the only process of animation. For to stop the breathing is to stop the living (LONGEVITY — Chapter 3). The products of cell disintegration are liquids and gases, of which the gases from the greater part. The major portion of the gases is eliminated via the lungs. The rest leaves the body through the kidneys and skin. The liquids also leave the body via the kidneys, lungs and skin.

Cells Are Not Produced By Food

We reiterate that what man eats and drinks does not produce cells or nourish them. The cells all come from the Parent Cell, the beginning point of the body. The body cells do not depend on what man eats and drinks. They are not nourished — that is another illusion. The cells are far above the nutritional level, are independent of nutrition, and are self-existent and eternal — a fact, demonstrated by Carrel.

Upon the demise of the body, the tissues, glands and organs disintegrate, setting free the cells that constitute these structures; and the immortal cells return to the Cosmic Ether, whence they Came. The cells are perpetual and indestructible, as are all primal elements of the Universe, to which order the body-cells belong.

Early Men Were Breatharians

These facts have constrained unorthodox physiologists to advance the theory that the first men in the early days were Breatharians, deriving from the Divine Breath of Life (Gen. 2:7) all the substance they required for its existence, and they lived from eighty thousand to a hundred thousand years; or until they were ready to return to their spiritual home, as stated in our work on LONGEVITY, Chapter 12.

The human body is now known to be a composition of trillions of cells. Each cell is a mass of millions of atoms, each of which is a globular system with "planets" whirling with tremendous speed. What use, asks the unorthodox physiologist, has such a system for food and drink? What use has our solar system for food and drink?

The Miraculous Cell

The body cells are endowed with unsuspected powers and astounding properties. Despite its smallness, the cell is an exceedingly complex organism that does not in the least resemble the favorite abstraction of the modern chemist — *a drop of gelatin surrounded by a semi-permeable membrane.* While the structural complexity of the living cell is disconcerting to the orthodox biologist, its chemical constitution is still more intricate. A droplet of water forms a tiny microcosm containing, in a state of extreme dilution a great variety of cosmic elements. The body cell is practically a duplicate thereof, but raised to the exalted plane of divine animation. Carrel says that the Parent Cell and the Cells that come there from are "not made of extraneous material, like a horse" (P. 107).

He means that the cells are not composed of what man eats and drinks. He states that while the body is composed of cells as a house is of brick, the body is born of a cell that makes more cells, just as though a house originated from one brick, "a magic brick that would set about manufacturing other bricks. Those bricks, without waiting for the architect's drawings or the coming of the bricklayers," says Carrel, "would assemble themselves and form a house" and all parts thereof.

Lesson No. 3
Physical Immortality

"The human frame as a machine is perfect. It contains within itself no marks by which we can possibly predict its decay. It is apparently intended to go on forever." - Monroe. John Gardner, M.D., in his work "Longevity," wrote: "Before the Flood men are said to have lived 500 and even 900 years. As a physiologist I can assert positively that there is no fact reached by science to contradict or render this as improbable. It is more difficult, on scientific grounds, to explain why man dies at all, than it is to believe in the duration of human life for a thousand years."

Dr. Foissac wrote in his book on "Longevity": "The long life of the biblical patriarchs is a fact more rational and more in accord with the known laws of physiology, than is the brief existence of the men who inhabit the earth today (Part II, Chapter 22). After years of investigation, the famous Metchnikoff declared that deterioration of the bodily structure and old age are due to minute quantities of poisonous substances in the blood.

Why Man Degenerates

In his book, "Prolongation of Life," Metchnikoff furnishes the first logical explanation in modern times of the degenerative changes occurring in the body, and why. His findings have been confirmed by leading researchers, including such noted doctors as Crile, Emphringham and Carrel. 1. Crile said, "There is no natural death. All deaths from so-called natural causes are merely the end-products of a progressive acid saturation." 2. Emphringham declares, "All creatures automatically poison themselves. Not TIME but these toxic products produce the senile changes that we call old age." 3. Carrel asserted, "The cell is immortal. It is merely the fluid in which it floats that degenerates. Renew this fluid at proper intervals, and give the cell proper nourishment upon which to feed, and, so far as we know, the pulsation of life may go on forever.... Quickly, involuntarily, the thought comes: Why not with man? Why not purge the body of the worn-out fluids, develop a similar technique for renewing them — and so win immortality?" (Man, The Unknown).

Carrel was great, but he missed some vital points. He could not rise above the medical theory of cell nutrition. We show in our work "THE NUTRITIONAL MYTH" that the cells of the body are self-sustaining, self-existent, eternal. They are far above the nourishment level.

In the attempt to nourish cells which need no nourishment, lies one of the errors that hurry man to the grave. Feeding increases the degenerative process, whereas fasting, the opposite course, institutes the opposite effect — the process of rejuvenation. The body is composed of trillions of cells. Carrel shows that all organs and structures that constitute the body as a whole, come from the original Parent Cell by a process of cell division and sub-division. That fact is definitely stated in all text-books on anatomy. It is highly important to note that Carrel further states that the Parent Cell and the Cells that come from it, are "not made of extraneous material like a house." He means that the cells are not made and composed of what man breathes, drinks and eats.

That being an admitted fact, then as the cells are not the product of food, they are independent of food and are not sustained by it. What food does not produce it cannot sustain. As the body is not composed of "extraneous material," we cannot feed it with such material, as such material is foreign to its constitution. The body is composed of cells, not of food; and the cells are composed of molecules, which are composed of atoms, which are composed of electrons, which are whirling centers of vibratory force in the ether. Recent observations show that the entire Material World is only a visible manifestation of varying wave forms. Dr. H. H. Sheldon, University of New York, wrote: "We live in a world of waves. The further we delve into the ultimate structure of Matter, the more obvious it is that nothing exists except in wave form. Electrons, long thought to be the ultimate particles of which all matter is formed, have now been shown to have a reality only as a wave form, while no atom consists of a bundle of such waves."

This means that the cells consist of a bundle of such waves, as cells are composed of nothing but atoms. This further means that the body consists of a bundle of such waves, as the body is composed of nothing but cells. So far as such physical structures as man and animals are concerned, this doctrine might be contested by the old school of scientists; but Sheldon says: "We as individuals undoubtedly have no existence in reality other than as waves, multitudinous and complicated centers, perhaps, in what we call the other.... We are analogous, in a sense, to the sounds that issue from a piano when a chord is struck, or when a symphony orchestra sounds." Electrons do not eat, atoms do not eat, molecules do not eat, cells do not eat, and the body does not eat.

Then why does man eat? That question science cannot answer; but we answer it according to Cosmic Law in our work, THE NUTRITIONAL MYTH." We said: "Cell and body nutrition is a myth. What man consumes does not supply cell nutrition by assimilation as taught by science. The ingested substances merely produce activity in cell function by stimulation and not by nutrition. Two types of stimulation seem essential for the function of living cells: Vital and Chemical. The nerves supply the Vital from the Life Source, and air, food and liquid supply the chemical. The ingested substances contact and stimulate the cells into action, and pass from the body through the eliminative channels, as flowing water turns the wheel of a mill, thus activating the machinery in the mill that does the grinding." - P. 17.

The Transportation System

Consider the water and the water wheel. The water below the wheel is the same that activated the wheel as it passed. The same substance that leaves the body as waste is the same substance that enters the body as air, liquid and food. In their passage through the body, as the water passing the wheel, they stimulate and activate the cells which compose the organs and tissues which constitute the body.

The substances man breathes, drinks and eats enter the body and pass into the blood. The blood, as a transportation system, carries the substances to the cells where they stimulate and activate the cells, but pass on without becoming a part of the cells — noting, the water passes on without becoming a part of the wheel or the mill. And as the water activates the wheel in passing, it activates all the machinery of the mill.

Carrel, as an orthodox medical doctor, went no farther than the cell and held to the medical theory of nutrition. His experiments showed the cell to be immortal. When the fluids fail to carry off what he regarded as the excrement of the cells, he found that the cells degenerate, grow senile, and show signs of dying. Each time they were rejuvenated by a cleaning process. Here again Carrel missed a vital point. Body cells do not die; and he himself declared them to be immortal. The cells simply sink below the life-level of vibratory action because their magnetic poles become corroded by the acids that are not carried off and eliminated by a clogged blood stream. Naturists know that patients recover quickly under fasting because fasting reduces the amount of pollution in the blood; thus, freeing the cells of the injurious effect of the acids. Crile said that "all deaths from so-called natural causes are merely the end-point of a progressive acid saturation." In fasting is the process of renewing the fluids mentioned by Carrel. That is also the process of rejuvenation as many experiments have shown. This startling knowledge leads physiologists to declare that decrepitude and physical death, except by ancient, are due to: (1) polluted blood system and the (2) accumulation of garbage.

This results in a corrosion of the magnetic poles of the cells, making them incompetent to receive the animating amount of vital vibrations of the Life Force, and they fall below the level of animation. Carrel called that condition the death of the cells.

If we keep the cells from dying, we keep the body from dying. For the body is composed of nothing but cells. Neither do the cells die. They are immoral. But when their vibratory rate falls below the life level rate, the body accordingly sinks below the life level plane: and that state is what we term physical death.

Blood Purification

Elimination therefore appears far more important than feeding. If blood purification by the kidneys were stopped, man would die in three to five days. If blood purification by the lungs were stopped, man would die in three to five minutes. In the matter of health and life we always come back to the blood. "It is merely the fluid in which the cell floats that degenerates," said Carrel, and he added, "Why not purge the body of the worn-out fluids, develop, a similar technique for renewing them — and so win immortality?" That simply seems to be the secret of physical immortality. With every poison that man can inhale, drink and eat which does not kill him instantly, his Vital Stream is changed from the River of Health and Life to the Pool of Pollution and Death.

In due time, after much suffering, man dies "of a progressive acid saturation" of his blood, tissues and cells, in fulfillment of the cosmic law that he reaps as he sows (Gal. 6:7).

Lesson No. 4
Vital Adjustment

Carrel and Lakhovsky are the greatest Scientists of the age so far as the constitution of Man is concerned. The former dug deeper into the body and its processes perhaps than anyone else in modern times, while the latter swept the universe in his search for the secret of Life.

Carrel devoted a chapter of his work, Man The Unknown, to the important subject of Adaptive Functions, and declared them to be responsible for the duration of man's life. He showed why man's body, composed of soft, alterable matter, susceptible of disintegration in few hours, lasts longer than if made of steel. He wrote: "Not only does he last, but he ceaselessly overcomes the difficulties and dangers of the outside world. He accommodates himself, much better than the other animals do, to the changing conditions of his environment. Such endurance is due to a very particular mode of activity of his tissues and humors. The body seems to mold itself on events. Instead of wearing out (collapsing) it changes. Our organs always improvise means of meeting every new situation, and these means are such that they tend to give man a maximum duration. The physiological processes... always incline in the direction leading to the longest survival of man. This strange function, this watchful automation with its specific characters, makes human existence possible. It is called Adoption."-191-2.

People know so little concerning the adaptive functions of the body. Hence, it will be difficult for the reader to understand that the weaker his body becomes the longer its duration under adverse conditions.

Carrel himself appeared not to know that man "persists in living despite physical, economic and social upheavals," because of the fact that as the "body seems to mold itself on events," it suffers in the process a corresponding weakness which will be explained as we proceed, and the explanation will be so new to the student that he will miss some vital points unless he reads this lesson several times.

Disease Germs

We shall begin by referring to the fraudulent claim that the world is filled with evil entities which attack healthy persons without reasons and the result is "disease," — something so dangerous that it must be combatted and cured. A deceived world believes in this and rejects evidence that exposes it. Carrel believed in it to the extent that he asserted, "scientific medicine has given to man artificial health" (P. 311).

It requires great prejudice to blind a scholar so completely that he cannot see the obvious absurdity of such a statement. According to cosmic law, man reaps as he sows; and that law has no exception (Gal. 6:7). We shall use that law as a guide to lead us through the wilderness and confusion created by Carrel's "scientific medicine."

Good Health Is Not Immunity

Doctors volubly discuss the theory of "resistance to disease," and Carrel believed in it. Supposedly good health makes man "immune" to disease — breeding influences, to the attack of germs, to vicious habits, to hostile environment. The theory of "scientific medicine" is that man is attacked by disease because of a weakening of the natural body defenses. Alfred Pulford, M.D., M.H.S., F.A.CT.S., a medical practitioner of fifty years, wrote: "The exciting and contributing causes of pneumonia may be, and are, legion; but they all simmer down to the one point, viz, the breaking down of the natural body defenses (Truth Teller, April 1944, P. 2).

In his daily column in the press of June 8th, 1944, Irving S. Cutter, M.D., said: "Now is the time to think of getting rid of that chronic winter cough. Yes, bacteria are responsible. But their very existence with all the irritation they can create, means that basic resistance (of the body) is too depleted to throw them off."

The better doctors know the germ theory of disease is false. If it were true, neither man nor beast could live long. They would be literally devoured by "disease germs." Natural science shows that correspondence must prevail as between the living organism and its environment. That is the positive, primal, and fundamental condition of Existence.

It was impossible for primitive man to come into physical being unless the proper, perfect and harmonious condition of his Environment prepared the way. Furthermore, the health condition of man's body can never be any better than the health condition of his Environment, with the condition of which his body must always be in correspondence.

Bananas grow not in a cold climate because their constitution does not correspond with such climate. Salt water fish does not live in fresh water because their constitution does not correspond with such water — that is natural science. Man is the most perfect of all creatures, and is able to rise superior to bananas and fish by reason of his body's "watchful automatism with its specific characters" which "make human existence possible." He can modify the condition of a hostile climate or adjust himself to it as to live for a limited time in the hottest and coldest regions on earth. The Law of Correspondence in this case means that a condition of harmony must exist as between living things and their environment, or they will die and disappear. That does occur occasionally, causing certain plants and animals to become extinct.

Spencer formulated the Law of Eternal Physical Existence and sought to show that Death was the end result of environmental changes which the living organism had not adapted changes to meet, thus creating a condition of discard and friction that sent the body down to death.

The human body is so perfectly constituted that it possesses the power to prolong its duration by adapting itself to conditions so adverse that they would otherwise cause not only early death, but instant death in some cases.

On this point Dr. Charles W. Greene wrote: "As the air exhaled from the lungs contains a large proportion of carbon dioxide and a small amount of organic matter, it is obvious that if the same air be breathed again and again, the proportion of carbon dioxide and organic matter in it will increase until it becomes decidedly unfit to breathe. It is remarkable fact that the organism, in time, adapts itself to a very vitiated atmosphere, and that a person soon comes to breathe, without sensible inconvenience, an atmosphere which, when he first enters it, feels intolerable. But such an adaptation can occur only at the expense of a depression of all vital functions, which must be injurious if long continued or often repeated" (P. 286). There is a definite statement of what occurs in the body as its "watchful automatism with its specific characters snake human existence possible" in an atmosphere so badly poisoned that it feels intolerable when one first enters it. The adaptation occurs *"at the expense of a depression of all the viral functions, which must be injurious if long-continued or often repeated."*

That is how the body builds up "basic resistance" to inimical influences and unhealthful conditions. We must first weaken the body's vital powers before it will submit without protest to "the dangers of the outside world" and the evil effects of bad habits.

Immunity Reduces Power To Resist

The secret of Vital Adaptation is the riddle that puzzles the doctors. They term it immunity. The body acquires immunity to dangers by reducing its powers to resist them. Exactly the opposite of what medical art teaches. We shall recite another instance of this weakened condition which medical art terms immunity. Greene continues: "This power of adaptation is well illustrated by an experiment of Claude Bernard. He showed that if a bird is placed under a bell-glass of such size that the air contained in it will permit the bird to live for three hours, and the bird is removed at the end of the second hour, when it could have survived another hour, and a fresh, healthy bird is put in its place, the latter will die at once" (Kirks Physiology, revised by Greene, P. 304).

According to medical theory of "vital resistance," the fresh healthy birds should have resisted the effect of the polluted air in the bell-glass and lived for three hours. But it died immediately; whereas, the other bird that had been under the bell-glass for two hours and suffered a certain degree of debilitation by reason of breathing the poisonous exhalations of its own body, could have lived for another hour.

This illustrates what Carrel means when he says, "Our organs always improvise means of meeting every new situation; and these means are such that they tend to give man (living creatures) a maximum duration. The physiological processes (of all animals) always incline in the direction leading to the longest survival of man."

Man in civilization is born into and grows up in an environment of polluted air that may kill in a day a wild Indian of the hills, whose vital body would react so violently to the shock, that death may soon result.

No matter how repugnant or destructive a thing may be, we can endure it, provided time is given to secure the efficient operation of the body's power of adjustment, whereby is prevented a violent swaying of vital activities from one extreme to the other. Only sudden and violent changes become immediately destructive to life, even sometimes when it is a change from evil to good habits. By the reduction of its vitality, the body pays in the process of becoming "immune" to discordant conditions, to poisonous substances, to evil habits.

That is the reason why the vigorous Indians of America became a "dying-race" when they came in contact with the enervating habits of the white man of Europe who survived in spite of the evil effects, because he was born in and grew up under those health destroying conditions. His body was accustomed to them, was adjusted to them; but the body of the healthy Indians was not, and they died like flies.

All evidence proves conclusively that the more vital the organism, the more quickly it succumbs to unhealthful conditions and harmful practices. That is another paradox. We live in a world of illusion.

There is a natural condition of Vital Adjustment to unhealthful conditions and harmful habits, but no Vital Resistance. That is another fallacy that belongs in the same category with the absurd theory of "contagious diseases." It is a paradox that the body, in a weakened condition, will tolerate and endure longer than a more vital body, the various evil practices and inimical influences which it cannot control, and which it must endure or die. Carrel says "the body perceives the remote as well as the near, the future as well as the present," and it prepares accordingly (P. 197). Creative Intelligence knows that the weakened body will endure more and live longer under adverse conditions. This appears paradoxical, yet the truth of it is, every scientist can demonstrate for himself.

Conditions That Destroy Health

The condition of the living organism is governed by habits, environment and climate. If these are good, vigorous health and longevity are the rewards. Conversely, when this trio of factors is bad, the body is forced to adjust itself to endure them, or perish. It is either endurance or death. But for the Law of Vital Adjustment, the race had perished ages ago. As it is, a gradual degenerative process through the ages has reduced man's life-span from eighty thousand years to an average of much less than eighty. It is Vital Adjustment, not Vital Resistance, that enables the body to survive for a few miserable years, with aches and pains, in air so foul that one in vigorous health would be in danger of dropping dead by suddenly coming in contact with it, as the vital bird thrust into the air that was polluted by the exhalations of the previous bird under the bell-glass.

Danger Of Smoking

The Law of Vital Adjustment makes it possible for the body to tolerate man's evil habits. That makes it possible for the smoker to endure and even enjoy his poisonous pipe. The same pipe would make a vigorous non-smoker ill, or might cause death, as death in such cases has been reported. The vital youth in his ignorance takes his first smoke. His vital body reacts with such vigor against the dangerous poison that sickness results. The degree of his sickness is the measure of his vitality. The more vital his body, the sicker he is.

Here are the poisons found in a chemical analysis of tobacco, nicotine, carbon monoxide, carbon dioxide, ammonia, methane, methylamine, hydrogen-sulphide, furfural, nicotelline, pyrrole, pyridine, picoline, lutidine, collidine, formaldehyde, carbolic acid, prussic acid, arsenic. Chemical analysis shows that cigarettes contain the following active poisons: furfural, acrolein, diethylene, glycol, carbon monoxide, pyridine, nicotine, carbolic acid, ammonia, and a host of tarry substances.

The tarry substances in tobacco smoke are approximately 2.25 percent. They stick to the walls of the lungs, forming a coating that obstructs the free passage of oxygen into the blood. Those who must breathe air laden with tobacco smoke are injured almost as much by it as are the smokers. These are the poisons to which the body of the smoker adapts itself in order to survive. It is either endurance or death.

The use of purgatives and laxatives in the case of constipation illustrates the work of the Law of Vital Adjustment. As these poisonous substances are used to force the bowels to move, the body slowly adjusts itself to them and from time to time the dosage must be increased to make the body act, or some other poison more powerful must be used.

As the horses grow weary drawing the heavy load, the whip must be used harder to drive them on. As the body weakens from constipation and the poisons that force bowel movements, the dosage must be stronger or more powerful poisons used to make the body act.

The youthful smoker disregards the warning reaction of his body. He continues the harmful habit, and gradually the prisons of the tobacco weaken the nerves and reduce the vitality to where it cannot fight back. Now he can smoke in comfort and experience immediately no ill effect. He has established "immunity' say some people. But in so doing he is taking the short-cut to the grave. The body slowly sinks into a process of degeneration and slow suicide under the power of the deadly enemy.

We may well say that what is termed the Law of Vital Adjustment is also the Law of Vital Reduction. It is the process of vital reduction that brings the body into subjection to any devitalizing substance, influence, habit or practice.

Toleration By The Body

The body's vitality is reduced to save it from sudden death and prolong its duration. Instead of dropping dead, one dies by inches, and in the process of slowly dying one suffers until the body can endure no more and the grave ends all. Great is the power of the body to adjust itself to conditions and poisons that would destroy it; far greater than man can imagine. The Law of Vital Reduction will enable the body to adjust itself to the point where the opium addict can take at one time a dose of dope so large that it had killed him quickly had he at first taken a dose that large.

The law may cause the body to adjust itself to the point where the venom of the reptile will fall to kill. The Grit of October 14th 1951, reported the case of Bill Haast who is said to be in that condition. The account is as follows — "He is full of snake poison himself, for when he first began handling the reptiles he set out to make himself immune to their bites. The serpentarium owner inoculated himself with larger and larger doses of snake poison and now believes he is the only man in the world genuinely immune to all type of snake venom. He has been bitten by nine cobras and many other times by rattlers, mocassins, corals, and various other kinds of killers."

As the Law of Vital Adjustment operates in the case of smoking and other bad habits, so has it operated in the case of eating. When man first began to eat, the substances that entered his then rudimentary stomach were poisonous to his body, as in the case of tobacco. He persisted in eating and the Law of Vital Adjustment brought his body into harmony with the habit.

But such on adaptation could occur only at the expense of depression of all the vital functions, which must be injurious if long continued or often repeated. That is the law and there is no exception to its operation. Some substances man eats are still poisonous to him after all the long ages he has followed the practice. His body has never been able to adjust itself completely to some of the things he eats.

Tea and coffee still make sick some who drink these, and no one can avoid their harmful effects. As the flesh eater becomes a vegetarian, the Law of Vital Adjustment goes into operation, and the time comes when even the odor of the meat market or of boiling beef, which was formerly so fragrant to him, becomes obnoxious to him. These are facts that are unbelievable in him who knows them not by actual experience.

Immunity

What medical art terms "immunity" is a condition produced by reducing the body's vitality. The reduction is accomplished by weakening and dulling the nervous system, and it may have far reaching effects.

The Federal Bureau of Census discloses the shocking fact that the number of persons 15 to 19 years old in this country in 1950 was 14 percent less than in 1940, whereas the population in the same period increased 19,500,000. The startling decrease in the number of young people may be the damaging effects of vaccination and inoculation — in an effort to make man disease-proof. Perhaps those who manage to survive the process and live long enough to reach adulthood, will never know the meaning of the term *good health*. The damaging effect of vaccinating and inoculating all school children seems to appear in the young men of draft age. The press of June 21st 1952, reported that 11.6 percent, or 1,443,315 of the 12,416,129 men it classified from 1948 up to April 30th under the draft system have been put to Class 4-F, as unfit for military service.

This includes the states, the District of Columbia and the territories. The Canal Zone, where the medical rule is less rigid in the vaccination and inoculation of children, had the lowest percentage of 4-F's.

Acute diseases are nothing more than symptoms of the body's reaction to internal poisons, or a condition of "acid saturation" as Crile calls it. There is no absolute immunity to internal poisoning. The body cannot be saturated with poisons and not suffer a corresponding degenerative change. The condition of apparent immunity to poison, to bad habits, or to so-called disease, arises because the poisonous substances, or the vaccines and serums, dull the nerves and reduce the body's vitality, making the body unable to react against the poisons that a more vital body would cast off in the eliminative process which are termed diseases. Man gains so-called immunity by paying the price. That price is a dulling of the nerves that reduce the body's vitality, and that is the short-cut to the cemetery.

    The wholesale practice of vaccination and inoculation is poisoning the body, reducing its vitality and decreasing its duration so fast that the death-rate after the age of 95 is steadily rising. Professor C. H. Forsyth of Dartmouth College was reported in the press of July 30th 1029, as stating: "The expectation of life from the age of 45 [onwards], is the lowest of which we have any record. Far lower than it was forty years ago — and it is still falling."

Lesson No. 5
Body Changes

"Man ceaselessly overcomes the difficulties and dangers of the outside world. He accommodates himself to the changing conditions of his environment. The body seems to mold itself on events. Instead of wearing out, it changes. Our organs always improvise means of meeting every new situation; and these means are such that they tend to give us a maximum duration. The physiological processes always incline in the direction of the longest survival." — Man, The Unknown.

These statements by one of the great scientists of the age, so far as living organisms are concerned, mean much more than many understand. By reason of the perfection of the body and of the power of vital adjustment possessed by it, a power little understood by the greatest scientist, man has been able to survive through the ages in spite of his changing environment and his evil habits. The toll taken through the ages by the evils and dangers thus thrust upon the body, has been a deterioration of the body's organs and structures that has reduced its duration from a period that probably once covered nearly a hundred thousand years to that of much less than one century in Modern times.

Misleading The Multitude

Modern institutions attempt to conceal the facts of human degeneration by deception, by the destruction of ancient records, and by leading people to believe that the race is progressing, moving upward, and that man now stands at the very pinnacle of his earthly career, that his scope of knowledge is greater than the race has ever before possessed.

There is not a scrap of evidence to support these claims. They are empty and misleading. They fall before every impartial investigation, and do much damage and no good. They weaken all desire to live a better life and discredit those who present the facts. They constrain the multitude to regard and treat as an enemy any teacher who would show the way back to the better life, making it dangerous to expose fraud and advocate race improvement. The elements of Time and Tear have no effect on the living body. The turning of the earth on its axis means nothing to a body that is repaired and renewed every minute of its existence. The body neither ages nor wears — but it weakens. The causes of the weakness include man's hostile environment, his harmful habits, and medical treatment.

Carrel seemed to disregard habits and medical treatment. He saw only "the difficulties and dangers of the outside world." He is right in holding that a hostile environment is enough to weaken the body and send it to the grave, but the difficulties and dangers of man's habits and medical treatment are often more injurious than those of the external world.

As a result of the inimical conditions to which it is subjected, and the injurious substances that enter its receiving chambers (air organs and stomach), the body changes instead of wearing out, as Carrel says.

The body changes may be briefly considered under three main heads, viz., 1) a change in the organs and glands ruled by the Law of Vital Adjustment 2) causing changes in their functions, and 3) a change in the quality of the cells and tissues.

By this count, the body gently sinks into a slow process of degeneration instead of dying suddenly, giving the body maximum duration under the circumstances, because it is able to change as Carrel says. 1. The changes in the body's organs and glands, under the Law of Vital Adjustment cause some of them to fall below par and, in time, lapse into a state of dormancy or semi-dormancy. The others must increase in size and function to compensate for this loss, putting a strain on them as the body struggles to survive under the handicaps it is forced to face. 2. The change in function resulting from the change in organs and glands are numerous and are for the worse. A condition of slow deterioration sets in, and its symptoms are what doctors are trained to term "disease." In time these changes appear as diabetes, Bright's disease, arthritis, neuritis, lumbago, rheumatism, and so on, through the entire list of diseases. 3. The change in the quality of the cells and tissues appear in that condition of weakness known as "old age," where and when the activity and elasticity of youth are succeeded by the slowness and stiffness of decrepitude.

Harmful Practices From Birth

The changes are not the work of Time and Tear, but of harmful habits, a hostile environment, and medical treatment which begins after the baby is born, with vaccination and inoculation to make the body "disease-proof," and is continued all through life. The body is poisoned and weakened at the start, begins its decline into degeneration immediately, and is never given an opportunity to recover.

As the organs and glands become dormant or semi-dormant, the body and its functions change in its adaptation to evil influences that reduce much of its integrity and efficiency, and it loses much of its spiritual capacity and faculty to function on the spiritual plane.

The evidence of this loss is most apparent in the brain, and proof of this loss is found in the institutions of civilization that are filled with the insane and feeble minded. It is common knowledge that civilized man is mentally only ten percent of what he should be, and the few on earth still sufficiently sane to see the light are those who are jailed and liquidated on charges of obstructing social progress by exposing the social pattern.

Man as a Breatharian received all his substance of the Cosmos directly through his Air (Spiritual) Organs, and thus functioned on the Spiritual Plane while dwelling in the Material World. His spiritual capacity and faculty to function on the Spiritual Plane failed in direct ratio as his body changed and became more material through the internal changes suffered by his organs as a result of the body's adapting itself to the evil influences which it could not control, and which it was forced to endure or die. In the perfect state of Breatharianism, man's body was free of the clogging and dulling substances of the material world; and it must be free again, of these damaging substances before it can return to its perfect state, when man was competent to explore with his mind the Spiritual Universe and possessed actual knowledge of the fact that he is of that Kingdom and has eternal life. But one should not court danger by trying to return to Breatharianism too quickly.

In his present, degenerate state, man's body is afflicted with all the cumulative effects of his harmful habits, his hostile environment, and the poisons introduced into his body by the doctors. That is the change, as Carrel called it, which we know as the shocking stage wherein man's physical organism sinks into the decrepit stage termed old age. He is taught to expect it, as his body wears out. On the contrary, the best doctors declare that the body does not and cannot wear out, and is so perfect, as a machine, that it should go on forever.

Fewer Centenarians

As evidence of the progress of degeneration, Royal S. Copeland, M.D., former New York Health Commissioner, said in the press of June 27th 1980: "Fifty years ago there was a population of a little more than fifty million people in the United States, 4,000 of whom were centenarians. At the present time, with more than double the population of fifty years ago, there are only 2,841 people who have reached the age of 100."Dr. John Harvey Kellogg wrote: "Civilized man is dying. This melancholy fact is recognized by all students of anthropology. Such eminent economists as Major Darwin, son of the famous Charles Darwin, and Professor C. B. Davenport of Carnegie Institution, consider the case hopeless, and believe that man will ultimately become extinct through degeneration." - Good Health, August 1930.

Max Heindel, an author of note, in writing on "Man's past evolution constitution and future development," attempts to show in his book under the "Science of Nutrition," that decrepitude arid death result from a change in the body's organs, structures and tissues because of what man eats and drinks. He ignores "the difficulties and dangers of the outside world," which Carrel saw, and failed to notice the wholesale manner in which the race is being poisoned. Heindel wrote: "There is a gradual increase in density and firmness of bones, tendons, cartilages, ligaments, tissues, membranes, the coverings and even the very substance of the stomach, liver, lungs and other organs. The joints become rigid and dry. They begin to crack and grate when they are moved, because the synovial fluid, which oils and softens them, is diminished in quantity and rendered too thick and glutinous to serve that purpose. The heart, the brain, and the entire muscular system, spinal cord, nerves, eyes, etc., partake of the same consolidating process, growing more and more rigid. Millions upon millions of the minute capillary vessels which ramify and spread like the branches of a tree throughout the entire body, gradually choke up and change into solid fiber, no longer pervious to blood.

"The larger blood vessels, both arteries and veins, indurate, lose their elasticity, grow smaller, and become incapable of carrying the required amount of blood. The fluids of the body thicken and become putrid, loaded with earthy matter. The skin withers and grows wrinkled and dry. The hair falls out for lack of oil. The teeth decay and drop out for lack of gelatin. The motor nerves begin to dry up and the body's movements become awkward and slow. The senses fail; the circulation of the blood is retarded; it stagnates and congeals in the vessels. More and more the body loses its former powers. Once elastic, healthy, alert, pliable, active and sensitive, it (changes and) becomes rigid, slow, and insensible. Finally, it dies of old age."

Medical art says the body wears out. Carrel says it changes. Given above is a good description of the more apparent changes. From the changes listed many others occur that are less noticeable and harder to describe. It is unnatural for the body thus to change; and the changes do not come of their own accord. The body battles against them, but they are forced upon the body by the persistence of a hostile environment and man's harmful habits.

The Body Fights Against Changes

The changes come slowly, gradually, steadily, while the body is constantly struggling against the various handicaps in its fight to live. The evidence of these struggles are the symptoms termed "disease." The cause of the changes are not in the body, but "in the difficulties and dangers of the outside world," and in the difficulties and dangers of man's evil habits.

Rudimentary Organs

The great Carrel, in his work, "Man The Unknown," wrote: "The body seems to mold itself on events. Instead of wearing out (dying), it changes" (degenerates) (P. 192). We have described some body changes in the foregoing pages, showing how and why the body sinks into degeneration and death. There is another phase of the subject to which we shall now refer. On page 197, Carrel said that the body's intelligence possesses both a prevision and a provision. It perceives the remote and the near; the future and the present, and provides, by definite changes, for such conditions and emergencies as its prevision shows that it must meet — or perish. Carrel failed, for some reason, to proceed from there and present some evidence of changes which the body makes and has made in order to survive under new conditions that prevailed not at the time when man first came into physical being. Had he done so, he would have uncovered some strange and startling things.

Could Carrel, as a doctor, see the present rudimentary organs and glands in the body, and not understand that they must have been useful and functional at some early period in man's life? These rudimentary structures must represent changes the body has made in its struggles to survive under adverse conditions. Could Carrel view the stomach and intestines in modern man and not realize they were rudimentary structures in the days when man was a complete breatharian?

With that evil disorder Constipation so prevalent that it is termed the "national disorder," and with few folks free of hemorrhoids (piles), of stomach and bowel troubles, with thousands suffering from appendicitis and many dying from appendectomy, could Carrel pass these glaring facts and not perceive that something must be wrong? How could he, as a doctor, fail to understand that all this misery and these disorders indicate a shifting of the body from its original course? Even a layman knows that if a machine fails to perform efficiently a certain work, that failure is evidence that the machine is asked to do what it was not made to do.

Supreme Intelligence equipped the body, in its physical beginning, with all the structures that it would ever need under all reasonable circumstances. It was made perfect and complete. But it seems that man has strayed even farther from the true path of life than was ever anticipated by an omniscient Creator. Modern man has the rudimentary breast of the female. In some cases, they are functional, and such men can nurse babies, as shown by Clements in his Science of Regeneration.

He also shows that the male glands of generation appear in rudimentary form in woman, and vice versa. These rudimentary structures represent changes that have occurred in the body through the ages because of changed conditions. As a Breatharian, man had all the organs, both functional and functionless, developed and undeveloped, that he would ever need as he drifted down the stream of degeneration.

He fell from the plane of perfection by becoming a drinker of fluid and a consumer of food; thus, creating unnatural wants and desires that have dragged him down to misery and despair. He had become extinct but for the prevision of an omniscient Creator, who provided him with rudimentary structures for just such emergencies. As the rudimentary organs were needed and commanded into use by new conditions, new environment, and new habits, they responded to the command and developed to a functional degree. Thus, the body changed instead of dying. Huxley and Darwin declared that the rudimentary organs in men and women are the remains of structures that have been better developed in an earlier state of human existence. They are the anatomical remains of what has been, and are used by modern biologists in tracing lines of descent with modification, and in determining probable ancestry.

Huxley said, "Either rudimentary or vestigial organs are of no use, in which case they should have disappeared; or they are of use, in which case they are arguments for 'telegony,' which means that they are of past and future service or purpose" (Anatomy of Invertebrates, P. 68).

Darwin made a deep study of this matter and he wrote: "Any complex organ in a rudimentary state is direct evidence of its once having been functional, and in order to discover the many transitional grades through which it has passed, we must look to very ancient forms which have long since become extinct. Rudimentary organs of now trifling importance, have probably been of high importance to an early progenitor, and after being perfected in a former period, have been transmitted in a more or less changed condition by modified descendants, until of slight or no use. In all species, or varieties, correlated variations play an important role, so that when any part has been modified or changed, other parts have necessarily been similarly affected or modified — and so viewing it, Nature may be said to have taken pains to reveal her scheme of modification by means of rudimentary organs, embryological and homologous structures, but we are too blind to understand the true meaning of them" (Variation of Species, pp. 14, 147, 178).

In Lesson No. 7 appears the story of a woman who is working back to Breatharianism. She said: *"I have passed the eating stage and could not eat even if I desired, as my digestive apparatus has changed considerably, and is now unable to handle any fibre at all."*

Because it has no work to do, this woman's alimentary tract is shrinking back to its original rudimentary state as it was before man began to eat. In the beginning the alimentary tract was rudimentary, as are the mammary glands now on man's breast, and it reverts to its original condition when man, by not eating, gives it a chance to change back.

These are some of the changes in the body to which Carrel referred but failed to describe. For the body, it either changes or perishes; it either adapts or expires. So the body meets the emergencies by making the necessary changes. But Carrel was too materialistic to use his Mind and find within the kingdom of God (Luke 17:21) the facts he needed to aid him in discovering and describing how and in what ways the body had changed. The body, being subject to changes, can change in practically all directions to meet many emergencies and survive, even the changes that decrease its efficiency and diminish its duration.

The body can change to meet adverse conditions, and it can also change to meet favorable conditions. These changes are possible because of organic and functional changes within the body. As the organs change, their functions must change to a corresponding degree.

To describe these changes definitely and in detail would require observations covering thousands of years. Some surprising changes can occur in one generation. As where a man turns into a woman and vice versa. If a man can change into a woman so completely as to become the mother of a baby, or a woman can change into a man and become the father of a child, as Clements shows in his Science of Regeneration, then it should not seem so surprising or impossible that a breatharian can change to a gluttarian and vice versa. As a man becomes a woman, as sexual changes occur because of organic and functional changes taking place within the body, so a breatharian becomes a gluttarian for the same reason. The sex glands of a body of nine or ten years of age are in a rudimentary stage as a rule and not competent to function in a productive degree. But, as they are commanded into use by the boy's habits, they respond and he becomes competent to produce offspring.

The press of July 15th 1951, reported the case of a ten-year-old girl in Picayune, Misisippi, who gave birth to a seven pound son. The doctor said the baby was "perfectly normal." When similar reports come out of India, we in the U.S.A. think it is terrible. When it occurs in our own country, it is winked at and forgotten. Prejudice is a powerful influence.

Lesson No. 6
Man's Natural Home

It stuns a man to tell him eating is not natural. Most men never heard of people who live without eating.

We show in our work, "The NUTRITIONAL MYTH," that eating is an acquired habit, like smoking. The sensation of hunger rises from certain stimulation of the alimentary tract. "Appetite comes with eating."

An advanced scholar writes that there was not an Ascent of Man but a Descent of Man, and this theory is supported by ancient records and legends. Man did not spring from the slime of the sea or from a worm in the ground, as science claims. He came from another planet or star, travelling to this planet in a space ship, now called "flying saucers," several of which have been seen since 1947, and some have landed on earth and dead men have been found in them.

This man did not eat, but subsisted on cosmic elements. He was a Breatharian, and assimilated sunshine and cosmic rays from the atmosphere of the earth to which he had come, and acclimatized himself to its atmosphere and his new environment. In that distant age man dwelt in high places where air is purest and highly charged with ozone and cosmic rays. This cosmic substance he inhaled, and it was termed the Breath of Life. By it his body was animated and sustained. In the high altitude the weather was perpetually cool, but his powerful vitality kept him comfortable. In that day, according to legend, man had a life-span of nearly a hundred thousand years. He did not know somatic death according to Bagget Irand, who said: "During that time it was common to find men and women who were thousands of years old. In fact, they did not know [somatic] death. They passed from one accomplishment to a higher attainment of life and its reality. They accepted Life's true source, and it released to them its boundless treasures in a never-ending stream of abundance." — Life &Teachings of the Masters of the Far East, Volume II. Long ages pass and the time came when man decided to use pure rain water. So he added liquid to his sustaining substance. This man was blonde in complexion, had sparkling blue eyes that resembled the color of the sky and hair of golden yellow that resembled the sunshine.

Prof. Hotema   *Man's Higher Consciousness*   *Lesson No.6*

The ancient Greeks had a tradition of the Hyperboreans who dwelt in the mountains in a land of perpetual sunshine and ate only fruit, but originally, like the gods from whom they descended, subsisted on air and sunshine. They were never ill and the duration of their life was a thousand years. The word Hyperborean means beyond or in the mountains. Man's traditional "Fall" occurred when he migrated to lower levels, where he found fruit growing and ate thereof — an event symbolized by eating the apple. Only after man descended to the low regions of the tropics, where he found fruit growing in abundance, did he become a consumer of food and darker in color.

Altitude Is Beneficial

Science shows that climate and altitude govern man. Each race harmonizes with its environment. In the high, cool regions, in the warmer middle regions, in the low hot regions, the type of people differ, but in each region they are basically similar. According to climate, altitude and the condition of the air, so is man. By these he is ruled, his constitution formed, and his habits shaped.

Regardless of where or how man lives, his body is basically composed of and sustained by cosmic rays, either directly or indirectly, in the form of minerals condensed from the rays after they strike the earth's atmosphere. It is for this reason that in high altitudes, where cosmic rays are stronger, the air contains more minerals to sustain the body, making it easier to subsist on cosmic rays at high altitudes than in low regions.

In low, warm regions people are languid, listless, of low vitality and poor health generally, regardless of how they live or the kind or amount of food they eat. The air of such regions lacks freshness and vigor; it contains too much carbon dioxide and too little oxygen and ozone. Also, the humid decomposing humus in the soil emits odors of acid decay that further weaken the body and shorten its duration. The worst air, speaking generally, is the stagnant, stifling, warm air in low regions of the temperate and tropic zones. In the latter region occurs the lowest human degeneration, and in some of these regions the average life-span is surprisingly short.

Languor, listlessness, weakness, and poor health come when the body cells are saturated with acids that disturb their mineral balance.

They lack the capacity to receive and register cosmic radiations properly. When the mineral deficiency advances far enough, the organic radio fails to function on the life level, and that state is termed "physical death."

Lesson No. 7
She Eats Nothing

In regard to eating Judith C. Churchill wrote: "When you overeat one day, you are hungrier the next. Huge meals stretch your stomach and throw your appetite out of proportion. Conversely, the less you eat the less you want. . . . After you become used to smaller food intake, you may wonder how you have previously eaten so much." - Readers Digest.

What man has done, man can always do. There is an ancient tradition to the effect that the first men did not eat, and a London lady is trying to prove it on herself. The London Sunday Chronicle of June 17th 1951, carried a picture of Mrs. Barbara Moore Pataleewa, of London, with her story that her "diet" consists of air, sunshine and an occasional glass of water. The account states: "A woman of 50 who looks like she was only 30 claimed yesterday that she hates food, has beaten old age and expects to live at least 150 years. She has set out to do it by giving up eating. "Twenty years ago she ate three normal meals a day. Slowly for 12 years she reduced her eating until she was keeping fit on one meal a day of grass, chickweed, clover, dandelion, and an occasional glass of fruit juice. Five years ago she switched entirely to juices and raw tomatoes, oranges, grasses and herbs. Now she drinks nothing but a glass of water flavored with a few drops of lemon juice to kill the taste of chlorine."

(NOTE-Killing the taste of chlorine in the water does not remove that poisonous substance from the water, and in time the cumulative effects of the poison will appear in some ailment if she continues drinking that kind of water. — Klamonti).

"She says, 'There is much more in sunlight and air than can be seen by the naked eye or by scientific instruments. The secret is to find the way to absorb that extra — that cosmic radiation — and turn it into food; that is what I have done. Every year she goes to Switzerland for the purer air and climbs the mountains on a diet of water from the streams, 'You see,' she explains, 'my body cells and blood have undergone a complete change in composition. I am impervious to heat, cold, hunger or fatigue." She continues: Winter or summer, even in Switzerland, I wear nothing but a short sleeved jumper and skirt. In cold weather people stare at me. But while they shiver in furs, I am warm. I am as strong as any man, and need only four or five hours' sleep for mental relaxation.

Because I have no toxins in my system, I am never ill. I had to advance gradually from vegetarianism to uncooked fruit and then to liquid food. Now I am struggling towards Cosmic Food. I have passed the eating stage and could not eat if I desired, as my digestive apparatus has changed considerably and is now unable to handle any fibre at all.

"Instead of thinking that my normal physical life will end in ten years, I am growing younger. With patience anyone can do the same. The tragedy is that eating is considered one of the pleasures of life. To stop eating is to experience discomfort while the body is adjusting itself to the new course. I now find the very smell of food disgusting."

Vegetarians find the smell of flesh (meat) disgusting. If they have been vegetarians long enough, to eat flesh would make them sick. We know by this and other experience that when the Breatharian first attempted to eat, it made him ill, as the first cigarette makes the youth ill. Eating still makes man ill while fasting restores health.

The statement that this woman is comfortable while thinly clad in cold weather while others shiver in furs, proves the correctness of the ancient tradition that ancient man dwelt in high regions where the weather was perpetually cool, but was kept comfortable by his powerful vitality.

She says that her body cells and blood have undergone a complete change in composition, making her impervious to heat, cold, hunger or fatigue. She further says that her digestive tract has changed considerably and is now unable to handle any fibre at all.

Survival Is Nature's Goal

The great Carrel devotes an entire chapter to the subject of Adaptive Functions in his work Man The Unknown, stating that the body seems to mold itself on events, and "instead of wearing out, it changes" (P. 192). He continues: "Our organs always improvise means of meeting every new situation; and these means are such that they tend to give us a magnesium duration. The physiological processes always incline in the direction leading to the longest survival of the individual."

We have an example of some of these changes occurring in the body of this woman. Her alimentary tract is shrinking to its original rudimentary state as it was before man began to eat. In the beginning the alimentary tract was rudimentary, as are the mammary glands now on man's breast, and it reverts to its original size when man gives it a chance by not eating. This woman is proving in her experiment that it is regular, under proper care, for the body to regenerate and return to its original perfect state. As degeneration is a fact, regeneration is a possibility.

Breatharianism To Gluttarianism

As a Breatharian, man's alimentary tract was rudimentary and his lung capacity was much larger than now. The lungs decreased in size as eating forced the development of the alimentary tract and reduced the capacity of the air organs, because eating reduced the body's need for "cosmic food." From its original state of Breatharinism, the body has gradually changed through the ages and declined to its present state of Gluttarianism. The Body has changed from a Superior Entity that was made to subsist of Cosmic Substance to an Inferior Entity that subsists largely on the gross products of materialistic substances.

The body never wears out, as is claimed. It changes says Carrel. The change to which Carrel refers is the natural way that the body sinks into degeneration from misuse and abuse, which includes bad environment and all of man's bad habits. As the body sinks in degeneration, its vitality is rendered too weak to fight to the death against dangerous and destructive conditions. Instead of going down in sudden death, the existence of the suffering body is prolonged.

It is a "change" under the Law of Vital Adjustment by which man escapes from early death for days of misery that are pitiful for the sufferer. As the body changes to adapt itself to the downward course, it must also change to adapt itself to the upward course, as shown in the case of this woman. As the body sinks in degeneration under abuse, so will it rise in regeneration under proper care.

Buried Six Months And Lives

The body is so plastic that it readily yields to man's desires and practices, whether good or bad. This is shown by the Yogis of the East who are reputed to have achieved almost unbelievable powers, by concentrated exercises and by systematic control of breathing.

Man's body is composed of soft, alterable substances, susceptible of surprising changes in function, making it last longer than if made of steel. Not only does the body last, but it ceaselessly overcomes the difficulties and dangers of its environment and of man's bad habits.

The press of July 26th carried the account of a Yogi of India, novice, who had established a record by living in a state of suspended animation for six months in a grave, without food or drink.

When he emerged from the grave, at Benares, his clothes were said to have been worn away and his body covered with white ants. By rigid body discipline he was said to have forced his beard to stop growing, and his whiskers were no longer than when he was interred.

The statements were made by Dr. B. L, Atreya, professor of philosophy in the Benares Hindu University and general secretary of the Indian Society of Physical Research. The yogi, practicing the art which consists in suppression of all mental activities, discipline of the body, control of involuntary muscles, and a few other weird things, lay in the grave from September 25th 1941, to March 21st 1942, the doctor said.

Spent Time In Cell

The yogi spent his time in a pit cell, reinforced with brick and cement. The day he emerged from the tomb a crowd of more than 100,000 persons was on hand, the doctor stated.

An opening was made in the outer enclosure and then an opening in the all-around closed cell, the first slab of stone was removed by selected persons, some of them Hindu professors and college teachers.

Said Dr. Atreya: "The yogi was already awake, and he raised his hand to indicate that fact. Then he was dressed in new clothes, His old clothes were partly worn away under the influence of the atmosphere inside the pit and partly eaten up by white ants, some of which were found collected over portions of his body. He was then brought out covered with blankets and placed on an easy chair on high platform, visible to all. The yogi looked just the same as when he entered the pit, Even the beard on his face had not grown. He insisted on walking about 25 yards from the place, but we did not allow him to do it for fear of his being crushed by the crowd which wanted to get close to him and touch his feet." — Grit, July 26th 1942. It appears that the Yogi know certain secrets of the body and its function which enable them to suppress all mental activities, control the involuntary muscles, and withdraw the senses from natural outward expression.

Lesson No. 8
Materialism

The Materialist missed the point when he wrote: "God formed man of the dust of the ground.... For dust thou art, and unto dust shall thou return." — (Gen 2: 7; 3: 19). As a block of ice is invisible cosmic substance materialized, so is the body of man.

Fasting experiments prove that vitality, heat, minerals, etc., come not from what man eats. They are cosmic rays that materialize as visible substance. As cosmic rays condense and become visible as matter, they lose none of their properties, one of which is vital force.

Science says that vitality and heat come from the combustion of carbon compounds in food, according to its calorie theory, which erroneously compares the living organism to a steam engine

Lakhovsky held that the living cell is an electro-magnetic entity, activated by cosmic rays, the source of its vitality. The cell's development is also directed by cosmic rays, which materialize in the form of body minerals and create the vital condition needed for the synthesis of atmospheric nitrogen into body protein, as well as atmospheric carbon dioxide into body fat. It appears that four-fifths of the air we inhale is nitrogen. The body synthesizes this substance to form protein, which occurs by its union with hydrogen in the alimentary tract. The body cannot use the nitrogen of protein foods. Practically all protein nitrogen that man eat is eliminated in the form of metabolic end-products.

Weight And Vitality Loss Due To Autointoxication

During a fast the body loses weight because it is toxic. There is a condition of autointoxication, and the internal toxins liberated cause decomposition of body protein and fat. If the organism were sufficiently pure, no condition of autointoxication and no loss of weight would occur during a fast. Inhale the terrible odor from the body of a person who has fasted eight of ten days and you will think his body is rotten.

When one's vitality decreases as one stops eating, it is due to auto-intoxication which then begins, and not to lack of food to supply energy. Food supplies no energy. Lakhovsky showed in his experiments that protein and other substances occurring in the body are converted from cosmic rays by the body's physiological processes. Organic growth and maintenance, says Lakhovsky, are the, work of cosmic rays. The living organism is a materialization of these rays. They are subtle streams of substance of ultra-electronic and materialize into grosser minerals as they strike the earth's atmosphere. So the body is a materialization of "cosmic food." He demonstrated this fact by keeping unicellular organisms in sealed test tubes, measuring the amount of iron they contained before and after a certain period of growth. He found the amount of iron increased as the cells multiplied, even though the test tubes were sealed.

The extra iron came from the cosmic rays to which the cells were attuned. They absorbed these cosmic rays at an iron rate of vibration and the rays materialized as iron atoms, showing that the cells of the body are maintained by cosmic rays. Babbit showed that sunlight is converted into minerals in the body, according to the spectral colors, each corresponding to different groups of minerals. All that food does is to furnish a certain type of stimulation. As we advance from vegetarianism to liquidarianism, and then on up to Breatharianism, we change from the grosser forms of stimulation to the finer forms. Instead of the body getting its stimulation from food and liquid, it gets its stimulation from the elements in the air.

When the scientific theory of Materialism exploded, material science exploded with it and its textbooks became obsolete. Professor J. S. Haldane, the great astronomer, said: "Materialism, once a plausible theory, is now the fatalistic creed of thousands (of physical scientists), but materialism is nothing better than a superstition, on the same level as a belief in witches and devils. The materialist theory is bankrupt."

With the discovery that atoms are composed of electrons and protons, and that these elements are merely whirling centers of force in the ether, material science saw its fundamental theories swept into oblivion. Physicists and chemists now know that all Matter is vibratory electro-magnetic activity. Matter is composed of units called atoms and atoms are composed of varying numbers and arrangements of electrons and protons, which are tiny centers of vibratory activity in the ether. Each of these vibratory centers possesses a magnetic polarity that is positive in the proton negative in the electron. All matter is fundamentally the same basic substance. The different properties which distinguish the various types of matter in the human body, such as proteins, carbohydrates, fats, etc., are, basically, nothing more than the differences in the number and arrangement of the protons and electrons.

The protons and electrons are held in their orbits in the atom and regulated as to the combinations they form, by the field of electro-magnetism generated by their rapid motions. In order to transform any given type of matter into another, as proteins into fats, it is necessary only that the vibratory frequency of the electro-magnetic activity composing the matter be altered appropriately. The transformation of matter is accomplished by exposing it to vibratory activity of the appropriate frequency, impelled by a force greater than the force impelling the vibrations of the substance to be transformed.

A simple illustration of this process appears in the transformation of ice into water and water into steam by exposing the substance to heat. The normal vibratory frequency of the substance composing the ice is increased by heating until the substance assumes a gaseous form, and the ice is transformed into invisible elements that float in the air.

Live Without Eating

The press of January 31st, 1981, said: "Authentic reports from Salisbury, South Rhodesia, state that Mrs. A. C. Walter, a noted singer, has been fasting 101 days, during which time she has consumed only two to three pints of cold and hot water daily. Last October she weighed 232 pounds, so she decided to fast. She has lost 63 pounds and says that she is in perfect health, goes out to parties, and carries on with her public singing."

Bernarr Macfadden of Physical Culture fame reported a case where he fasted a man for 90 days. He wrote: "The man lost 75 pounds during this period. He weighed 300 pounds when he began this fast, and 225 pounds when the fast ended."

Macfadden adds: "If a bear can fast all winter, there is no reason why a man could not do the same thing."

The Great Body Normalizer

No measures known will so surely, safely and speedily normalize a deranged body as will fasting. It is the most natural and certain of all remedial procedures, it stops at once the introduction into the body of all new material, except air and water, thus releasing the organs from the labor imposed by eating, and giving them an opportunity to purge the body of the internal poisons responsible for illness.

On April 28th, 1929, Paul Urban, a German world war veteran and professional nurse, ended a 64 day fast, during which time he took a pint and a half of pure water daily. He weighed 165 pounds and dropped to 113 pounds. He stated that fasting rejuvenates the body and makes man live longer. He was 46. In the press of July 19th, 1929, appeared an announcement of cancer being "cured" by fasting, with a picture of the patient and his nurse, under which was this statement: "Albert Schaal, age 58, shown as the flax king of Manitoba, Canada, after a fast of 49 days under the direction of Dr. Harry C. Bond of San Francisco, is said by doctor to be cured of cancer."

In his "Believe it or Not," in the press of January 25th, 1938, Robert Ripley stated that for ten years Giovanni Succi travelled through Europe living exhibitions of fasting. His exhibitions, rigidly controlled, extended for periods of 30 to 40 days. During that time, he was in the public eye day and night. Included were 80 periods of 80 days of fasting, and 20 periods of 40 days of fasting — a total of 3200 days without eating, or eight years and 280 days without food in ten years. In his "Believe it or Not," in the press of January 16th, 1934, Ripley stated that Jekisiel Laib, of Grodno, Poland, fasted six days week for 80 years. Each Saturday he ate bread and water. His health was good. According to the dietetic experts he should have died of "mineral starvation."

According to the press of July 26th, 1942, a Yogi at Benares, India, was buried in a grave for six months without food or water. (Report given in this lesson). The press of November 30th, 1934, reported the case of a Jain priest, Muni Shri Mierilalji, of Bombay, who fasted for 259 days, taking nothing but water. He ended his fast in the presence of 500 co-religionists. The press of October 12th, 1948, reported the case of a British girl of 12 years who fasted for 18 months, taking nothing but water. The press of February 6th, 1937, quoted Mrs. Martha Nasch, age 44 of St. Paul, Minnesota, as asserting that for seven years she had eaten nothing, and affirmed her willingness to submit to surveillance to prove her claim. The press of May 31st, 1948, reported the case of a Chinese girl who had eaten nothing for nine years.

The case was reported to Dr. T. Y, Gan, of Chungking Municipal Hospital, and he went to see her. Her name was Yang Mei, she was 20 years old, weighed about 85 pounds, and led a perfectly normal life, except for not eating, and drinking very little water, she showed no signs of starvation, and appeared no different from other girls, Gan said, "I found it difficult to believe her story." The girl was never hungry, had no desire for food, and never asked for any. When asked as to why she did not drink more, she said that it made her feel uncomfortable. Her alimentary tract was so dormant and rudimentary that it could not take water without bad reaction.

In an article entitled "Forty Years Without Food," N. P. Chose wrote: "Caribala Dassi, sister of Babu Lamboxar Dey, a practicing pleader of Purulia, has been living for the last forty years without taking any food, not even water, and has been doing her regular household duties with no apparent injury to her health. Many respectable persons can testify to the truth of this statement. "-India's Message, January 1932. According to the press of May 27th 1987, Sirmati Bala, of Bankura, India, age 68, had touched neither food nor water since she was 12 years old.

One case is sufficient to show what is possible in a million other cases. Biologists are being convinced that eating is an acquired habit, like smoking, and a pleasurable indulgence rather than a physiological necessity. It is said that in India certain sects of yogis live without eating, and in the Himalayas there are many who consume no physical food.

Lesson No. 9
Body Building Material

All textbooks on anatomy teach that the human body is a composition of trillions of cells. The cells are not composed of food.

Science admits that the Parent Cell is not the product of food. It also admits that all the subsequent cells are not the product of food. It is law that what food does not and cannot produce, it does not and cannot preserve and sustain. To speak or write of cell nutrition and body nourishment is not only unscientific, but an admission per se of anatomical and physiological ignorance, even though the statements may be those of a great doctor. When health officers die as they do in their fifties and sixties, it shows that they do not know what they should.

One author says that we have one foot mired in an antiquated medical system that is dying, and with the other foot we are holding down a modern health system that is struggling to be born, which advocates Health by Healthful Living in harmony with God's Law of Life. The origin and work of the Parent Cell is a mystery; that is the cell which begins the building of the body. That cell comes not from the parents.

The so-called seed of the parents is not seed in the sense that it produces man. What is considered as the seed appears to do nothing more than to form a central, electro-magnetic point, around which occurs a condensation of invisible substance from the Cosmic Reservoir. The food one eats neither form nor sustain the body cells. Nor do they come from the, Parent Cell, which is just an electromagnetic center of crystallization and materialization, a pattern, around which Cosmic Rays materialize into cells that form the growing body.

In referring to this mystery, the great Carrel wrote: "The body builds itself by techniques very foreign to the human mind. It is not made of extraneous (foreign) material, like a house. It is composed of cells, as a house is of bricks. But it is born from a cell, as if the house originated from one brick — a magic brick that would begin making other bricks (of material that seemed to come from nowhere). Those bricks without waiting for the architect's drawings or the coming of the bricklayers would assemble themselves and form a complete house... as does the body and all its various parts" (Man The Unknown).

Whence come the cells? or the materials of which they are composed? They rise as shadows and become substance as the result of the condensation and materialization of Cosmic Rays. That substance is not food. It constitutes the elements of the Universe that have always existed and are eternal. As the so-called seed of the parents come near to each other, certain elements of each stand out separately and, coming nearer, these separate, individual particles merge and fuse as it were into each other, producing a clear field in which nothing appears. Finally, after a period of seeming quiescence, granulation occurs at a point between the places occupied by the gametes of the parents when merged, fused and disappeared from sight (P. 194).

When the so-called seed of the parents meet and fuse, they thus create a condition or electro-magnetic center, which is necessary for the occurrence of the phenomenon that produces man's body. That center attracts cosmic rays of a definite frequency, corresponding to the chemistry of that center. The rays crystallize around the center in the form of similar substance, and man comes into physical being under a magic process of transformation of invisible elements into visible form.

The great Carrel missed the point when he said that the body is not built of "extraneous material, like a house." The infant body must receive for its growth material from some extraneous source. That source is the invisible cosmic rays. The process of growth is the work of these rays as they materialize into blood, bone and flesh.

Discovery Amazed Material Science

A block of ice represents a materialization of cosmic vapor. So the body of man represents a materialization of cosmic rays. The one process is as simple, as complex and as mysterious as the other. Both processes are ruled by the same cosmic law. The scientific theory of cell reproduction cannot explain how the Parent Cell produces cells that form blood, bones, muscles, nerves, heart, brain and other organs. These differentiated cells each represent cosmic rays of different wave length.

The cosmic rays become visible by condensation and materialization of Invisible Elements that exist only as vibratory waves, whirling centers of force, termed electrons, concerning which Dr. H. H Sheldon, University of New York, wrote: "Electrons, long regarded as the ultimate particles of which all Matter is formed, have now been shown to have a reality only as a wave form, while an atom consists of a bundle of such waves." This discovery amazed material science and exploded the basic theories of Materialism. For it shows that the cells of the body consist of bundles of waves, not of assimilated food. So man is not what he eats. The body cells are composed of atoms only, and the body is constituted of nothing but bundles of vibratory waves.

So far as man's physical structure may be concerned, this doctrine might be contested by the old school scientists, but Sheldon says: "We as individuals undoubtedly have no existence in reality other than as waves, multitudinous and complicated centers, perhaps, in what we call the ether. We are analogous, in a sense, to the sounds that issue from a piano when a chord is struck, or when a symphony orchestra sounds."

This raises the question, WHY DOES MAN EAT? In our work 'THE NUTRITIONAL MYTH" we wrote: "Cell and body nutrition is a myth. What man consumes as food does not supply cell nutrition by assimilation as science teaches. The ingested substance merely produces activity in cell function by STIMULATION and not by nutrition. Two types of stimulation seem essential for the function of living cells: Vital and Chemical. The Vital is from the Source of Life, while air, liquid and food supply the chemical. The ingested substances contact and stimulate the cells into certain activity, and pass from the body through the eliminative channels, as flowing water turns the wheel of a mill, activating the mill machinery that does the grinding "-P. 17.

The process of organic growth from the electro-magnetic center formed by the Parent Cell, and the so-called process of nutrition, are one and the same phenomenon. Cosmic rays strike the chromosomes of our cells, which act as minute receptors of cosmic radiation, and the rays materialize in our cells into various chemical elements requisite for organic growth and maintenance. The magnetic chromosomes of the cells attract the electronic rays of corresponding vibratory frequency and they materialize in the cells as minerals.

What we term minerals are the foundation of the living cell. These minerals are substances that are universal in existence and eternal in duration. These substances are electrically charged particles of various minerals. The body is a complex of minerals, consisting of electrons combined into atoms and molecules. It is not composed fundamentally of proteins, carbohydrates and fats. The substances in the body are actually atoms of nitrogen, oxygen, hydrogen and carbon dioxide in various combinations, as water is a combination of hydrogen and oxygen. The nature of the substance depends upon its atomic combination.

Look To This Day

For it is Life, the very Life of Life.
In its brief course lie all the Verities
and
Realities of your existence;
The Bliss of Growth;
The Glory of Action;
The Splendour of Beauty;
For Yesterday is but a Dream,
And To-morrow is only a Vision;
But To-day well lived makes every
Yesterday a Dream of Happiness, and
Every To-morrow a Vision of Hope.
Look well, therefore, to This Day!
-From the Sanscrit.

"A thousand years hence the contents of this work will be as up-to-date as at this hour . . . writings and methods of living based on Cosmic Law are always in order and never become obsolete."

Lesson No. 10
The Aging Process

"By old age of the body, that does not age; by the death of the body, that is not killed. It is thy Self free from sin, free from old age, from death and grief, from hunger and thirst." — Chandogya Upanishad.

People want to look younger and live longer. To that end many books have been written and the authors died young as proof that they were incompetent teachers. Why does man grow old? If the earth's turning on its axis does not produce "old age," what does?

An unusual account of Old Age appeared in the press of July 19th 1952. The item, date-lined Chicago, said: "A four-year-old girl, who weighs only 7 pounds is dying of old age at University of Illinois Research and Educational Hospital. The child is a victim of progeria or premature senility. Doctors at the Hospital said it is one of the rarest ailments. Both its case and cure are unknown. The child, named Linda, entered the hospital January 28th 1948, when she was two months old, has been there ever since. A hospital spokesman said Linda is withered and wizened with thin, balding hair, is only two feet long and wears doll dresses and shoes."

Nothing happens by chance. We may not always understand the working of the law because our view of life is so limited. If our thoughts penetrated beneath the surface, we would find a cause for every effect.

We are not told why the child was taken to hospital when only two months old, but the aging condition of her body is evidence to prove the harmful effects an artificial mode of living. Everything natural is banished from hospitals and the laws that rule natural phenomena receive no attention in such places. In the matter eating, Abbe N. De Montfaucon De Villars states that the Ancient Masters ate food only for pleasure, and never of necessity (Compte De Gabalis, P. 63).

We are told that in the Breatharian Age man's in its perfection, required not that kind of stimulation which physical food now furnishes.

Our fundamental concept of man should be that, as we know him, he is a degenerate representative of the original. His environment, greatly changed and adversely affected by the conditions called civilization, and his habits and practices, most of them bad, have forced the body to alter its functions in order to survive. Otherwise it had perished.

In the course of long ages the body's functions have changed, by continuous adjustment, and developed a dependency upon certain kinds of stimulation, rising from man's environment and his eating and drinking habits, that were foreign to the body in its original state, when it received directly from the Air, the cosmic reservoir of all things, the stimulation needed to activate its cells. Poverty and wants are conditions created by man's living an artificial life. The less we need the more complete we are, and we attain perfection only when free of all wants. The more wants we have, the less complete we are, and the farther we incline from perfection.

Conditions Of Artificial Life

Daily experience proves that the body still continues to adjust itself to man's additional errors, such as smoking, drinking, and eating certain things. Some find it impossible to smoke, while others cannot tolerate certain foods which some seem to enjoy. So the body has been forced, by long ages of eating, either to adjust itself to the foreign substances man eats, or perish. Instead of dying quickly as a result of man's errors, the body changes and sinks into degeneration. Our organs always improvise means of meeting every new situation, and these means are such that they tend to give us a maximum duration under the circumstances. The functional processes always incline in the direction leading to the longest survival of man (Carrel, P. 192).

So-called food is foreign to the body. None of it enters into the body's constitution and construction. If man's body were built of what he eats, a process of physical transformation would in time change the body literally to resemble physically the things man eats. If man is what he eats, if the body were built of the food man consumes, the eating of pork would in time transform him physically into a pig. The body was forced to adjust itself to what man eats in order to survive. It was either adjustment or death. The adjustment has become so complete that man now seems to "starve to death" when deprived of that stimulation which food furnishes.

Return Must Be Slow And Gradual

The return or transformation to Breatharianism, where food is no longer essential for body stimulation, must be slow and gradual. Man must slowly reduce the amount of food ingested daily in order to give the body time to meet the new condition and adjust itself to the perfect physical state of long ago, when the air man inhaled supplied all the stimulation the body needed. We must also leave the polluted air of civilization, or perish.

Some scholars assert that man has been on earth six to eight million years, and produce certain evidence to prove it. Dr. W. C. Pei, research fellow of the Chinese National Geological survey, unearthed "The Peking Man's remains near Peipine in 1929," and this discovery, according to science, pushes back man's appearance on earth fifty million years.

Antiquity Of Man

In his book titled "Sree Krishna," Premanand Bharati states that this is the 28th Divine Cycle of which the first three sections, viz., Golden Age, the Silver Age, and the Copper Age, have passed away. We are now in the early part of the fourth section, the Kali or Iron (Dark) Age.

The Divine Cycle is composed of 12,000 Divine Years, each of which is equal to 860 human years. So that 12,000 Divine Years multiplied by 860 give us 4,320,000 human years, which is the length of a Divine Cycle, as this is the 28th Divine Cycle, that would be a total of 120,960,000 years. The Hindu scriptures state that man had been on earth 4,000,000 years when the Great Deluge occurred.

The author of "Sree Krishna" says: "These men (of the Golden Age) required little material nutrition; they ate very little food, consisting of fruit only, and drank water — and these between long intervals."

Some physiologists hold that it requires three-fourths of man's time on earth to descend from Breatharianism to Gluttarianism, which was accomplished by an alteration of the body's functions and needs as it adapted itself to the new conditions and practices with which it came in contact. This would place the Breatharian state so far back in the night of time that little evidence of it could be found, other than what we learn now by fasting a man, who begins at once to regain health when given no food, and even shows signs of growing younger.

Professor Morgulis wrote: "The acuity of the senses is increased by fasting, and at the end of his 31 days abstinence from food, Professor Levanzin could see twice as far as he could when his fast began."

Bernarr Macfadden of Physical Culture fame wrote, "I have consistently maintained that the body can be revived and rejuvenated in every way, mentally, physically, etc., by fasting."

Dr. Moeller said, "Fasting is the only natural evolutionary method whereby, through a systematic cleansing, the body can restore its equilibrium by degrees to physiological normality."

Mayer, eminent German physician, declared: "Fasting is the most efficient means known for correcting disease" (The Wonder Cure).

Dr. Densmore wrote: "We find one great cause that accounts for the majority cases of longevity — moderation in the amount of food eaten" (P. 295). Dr. Evens said: "Among instances of longevity, we have the ancient Britons, who, according to Plutarch, 'only begin to grow old at 210'. Their food consists almost exclusively of acorns, berries and water."

Drs. Carlson and Kunde, University of Chicago, found that a fast of 15 days restored the tissues of a man of 40 to the physiological condition of those of a youth of 17. This amazing discovery seems to explain the biblical statements, "His flesh shall be fresh as a child's; he shall return to the days of his youth (Job 33:25). And thy youth shall be renewed like the Eagle's" (Ps. 103:5). Here is more evidence of damage to the body by food. Eat little of a simple diet and we live longer and look younger.

Stop eating and the sick man automatically and spontaneously begins to recover his equilibrium. As if by magic the disorders disappear and health returns. For that reason fasting is often feared and bitterly condemned by some. A world of health would put some people out of business. The press of February 26th 1948, stated that illness brings physicians of the U.S.A. $1,500,000 daily."

How To Reverse Physical Appearance Of Aging

The turning of the earth on its axis has no affect on the human body. Looking older as the years pass is the effect of physical adjustment to adverse conditions, and shows more complete adaptation of the body to bad habits and bad environment. Drugs, medicines, vaccines, serums and tonics are not the answer. Supply better living conditions and the physiological process of degeneration changes to regeneration, and the physical appearance of aging will reverse.

All things that damage the body age the body. If food damages the body it ages the body. Sickness begins to age the body in childhood because sickness results from damage to the body. Drugs, medicines, serums, hot and cold baths used in sickness, age the body because they damage it. Polluted air, bad water, hard water, chlorinated water, tobacco, liquor, heavy manual labor, excessive exposure to the hot summer all kinds of riotous living — these age the body, and the body improves when such practice or habit ends. Remove the CAUSE and you have found the CURE. A library of medical books is unnecessary to teach that simple law of Cause and Effect. Under no circumstances can man stop breathing and live. Every living thing must have air or die. To stop the breathing is to stop the living. Water comes next. Man may go without water for days and live. If the air is very moist, he can live longer without water than in a drier atmosphere. Men at sea, when shipwrecked and have no fresh water to drink, are able to supply the body's needs by submerging the body in the water. The salt is filtered out as the water is absorbed through the skin.

No one can breathe too much good air or drink too much good water. But one can easily eat too much of the best food and the result is always bad. No eating habit is harder on the body or causes it to age faster than that of salting food. Anyone who claims the body needs common table salt should try drinking sea water when thirsty.

Constipation The National Disorder

The practice of eating being considered as bad, is further shown by the fact that few people are free of stomach and bowel troubles, while constipation is so universal that it is termed the national disorder. This should not be if eating were natural for man. If food were not foreign to the body, it should not derange the so-called food organs, the digestive tract and its accessories. If food were necessary to sustain the body, then the fasting of patients would be dangerous and could not be the "cure all" that it seems to be. We know the body adjusts itself to many abuses and becomes accustomed to many new conditions. So it can adjust itself to feeding. But little does man know that *all such adjustments can occur only at the expense of a depression of the vital functions, which must be injurious if long-continued or often repeated. That is what the body has suffered because of eating.*

While it appears from the evidence that eating is not natural, the body as a machine is so perfect that it can take such abuse and survive for a century or more, provided the amount of food consumed is not too great.

That fact was shown in the case of Ludovico Cornaro, who was a physical wreck at the age of 40 and told by his physicians that he could not live. He fooled them by turning to Nature and recovering health to such extent that he lived to be 103. Cornaro found that a simple diet of 12 ounces of solid food and 16 ounces of fresh fruit juices daily was comparatively better for him. On his 78th birthday his friends urged him to increase his ration. Reluctantly he agreed to an increase of only two ounces of the same food. In twelve days, he was ill with fever and pains in his right side. He returned at once to the 12 oz. ration, but suffered for 35 days. That was his only illness in 63 years on his frugal fare.

One case is sufficient to show what is possible in millions of other cases. Cornaro proved on himself the virtues of frugal feeding, contrary to medical advice that man must be "well nourished." The consumption of much food to build up the body's "resistance to disease," works the other way. Looking older, old age, a state of decrepitude that appears with the years, is the result of the body's adjustment to harmful habits and adverse environmental conditions. No hunter ever found a wild animal that showed signs of old age. Supply better living conditions, discard bad habits, move to an environment of good air, keep the home well aired, practice chastity, and the appearance of "aging" will retard and reverse in time to that where the body was when its equilibrium began to show imbalance or deterioration. That takes time, as the "aging" process must first come to a full stop before improvement can begin.

Lesson No. 11
Does Man Starve

"As long as we confine ourselves to the world of observation, we must continue in a state of bewilderment." — Robert Walter, M.D.

We live in a world of illusion. We are victims and prisoners of our five senses, and they are unreliable. We are not surrounded by what we think we are, nor do we actually see what we think we see.

The press of July 23rd 1951, quotes Dr. Theron Alexander, psychologist, Florida State University, as declaring: "The familiar saying of 'seeing is believing' is being seriously questioned these days. It is more the reverse; we tend to see what we want to see."

There are two systems of Thought — the backward and inward, and the forward and outward, the inductive and deductive, the empirical and logical. Both systems are claimed to be based upon Fact, but not upon the same class of Fact. One bases all practice upon the facts of observation, and claims that its processes are inductive. This is the backward and inward process, and Francis Bacon was its great representative. The other system, the forward and outward, is based upon a fundamental truth or principle, and was followed by the Ancient Masters, who taught that the secrets of the Universe are discovered by studying Causes instead of Effects (Rom, 1:20). All observable facts are the effects of something preceding; the something which preceded is the Cause, which, being discovered, constitutes an eternal verity which changes not, and so becomes the unchanging basis from which logical reasoning may be conducted.

Professor Robley Dunglison, one of the ablest authors and professors, warns us against reliance on observation, and quotes the man who —

"Saw with his own eyes the Moon was round,
Was equally sure the Earth was square,
For he'd travelled twenty miles and found
No sign that it was circular anywhere."

Dunglison recited many facts to show the fallacies of medical observation and the absurdities of medical practice that have grown out

Man's Higher Consciousness

Professor Hilton Hotema

© Frontline Distribution Int'l Inc. & Research Associates School Times Publication 2017

All Rights Reserved ©
No part of this publication may be reproduced, stored in a Retrieval system, or transmitted, in any form or by any means, Electronic, Mechanic, Photocopying, Recording, or otherwise, Without prior permission of FRONTLINE DISTRIBUTION INTERNATIONAL INC.

ISBN #: 9781683650195
LOC#: 2016940730

2017 Edition Editor: Prizgar G. & Nastasia Grant-Terrier
Print Coordinator: Prizgar G.
Book Cover Design: Ras Tzaddi Wadadah II

Table Of Contents

Prologue	i
Introduction	v

Lesson No.1
Physical Perfection	01
Breath of Life	01
Stages of Degeneration	02
Magnetism	02
Spiritual Potentiality to Physical Actuality	04
The Way to Improve Physical Man	05
So Called Civilized View	05
Why the Truth is Suppressed	06
Economic Freedom	06

Lesson No.2
The Living Cell	07
Chiropractic Law of Physiology	08
Where Did The Living Cell Come From	08
Is Eating Necessary	09
Elimination	10
Cells Are Not Produced By Food	11
Early Men Were Breatharians	11
The Miraculous Cell	11

Lesson No.3
Physical Immortality	13
Why Man Degenerates	13
The Transportation System	15
Blood Purification	16

Lesson No.4
Vital Adjustment	17
Disease Germs	17
Good Health Is Not Immunity	18
Immunity Reduces Power to Resist	20
Conditions That Destroy Health	21
Danger of Smoking	22
Toleration by the Body	23
Immunity	24

Lesson No.5
- Body Changes — 26
- Misleading the Multitude — 26
- Harmful Practices from Birth — 28
- Fewer Centenarians — 29
- The Body Fights Against Changes — 30
- Rudimentary Organs — 30

Lesson No.6
- Man's Natural Home — 34
- Altitude is Beneficial — 35

Lesson No.7
- She Eats Nothing — 37
- Survival is Nature's Goal — 38
- Breatharianism to Gluttarianism — 39
- Buried Six Months and Lives — 39
- Spent Time in Cell — 40

Lesson No.8
- Materialism — 41
- Weight and Vitality Loss Due to Autointoxication — 41
- Live Without Eating — 43
- The Great Body Normalizer — 43

Lesson No.9
- Body Building Material — 46
- Discovery Amazed Material Science — 47

Lesson No.10
- The Aging Process — 50
- Condition of Artificial Life — 51
- Return Must Be Slow and Gradual — 52
- Antiquity of Man — 52
- How to Reverse Physical Appearance of Aging — 53
- Constipation the National Disorder — 54

Lesson No.11
- Does Man Starve — 56
- Finding the Truth — 57
- Food Stimulates — 57
- It is the Body That Acts — 58
- People Who Crave Poison — 59
- Power of Adaptability — 59
- Fish Does Not Give Brains — 61

Lesson No.12
- Danger of Abrupt Changes 63
- Men of Great Stature 63
- Stature Originally Gigantic 65
- Man's Body Resembles the Planetary Bodies 66
- Misleading Reports 67
- Body Craves Food As It Does Poison 68
- Man Eats to Die 69

Lesson No.13
- Chronic Auto-Intoxication 71
- Body Tries to Maintain Balance 72
- Vitality Increases 73
- Eat Little And Live Long 74

Lesson No.14
- Body Needs 76
- Minerals From Cosmic Rays 77
- Sensation of Hunger 78
- Eating Is A Vicious Circle 78
- Atomic Energy 80

Lesson No.15
- Eating Poisons 81
- Dangerous Narcotic From Juice of Poppy 82
- Mice Unable to Live on Human Diet 83
- Evolution and Devolution 85
- Dormant Organs Ready When Needed 86

Lesson No.16
- Vegetarianism Is Bad 88
- Body Vitality Reduced 88
- Most Vegetables Are Not Natural 89
- Cereals are Bad Foods 91
- Fruits Easier Produced with Less Labor 92
- Alimentation and Decrepitude 93
- Earthy Salts Cause Old Age 94
- Fruits Have Little Earthy Matter 95
- Fresh Fruit 96

Lesson No.17
- Carnivorism Is Bad 97
- Butter, Milk and Cheese Less Harmful 98
- Reason For Increased Vitality 100

Flesh Foods Putrefy	101
Mode of Living Builds Cravings, Aches and Pains	102
48 Million Have Trichinosis	103
Table Salt	104
Opinions on Salt Eating	105
Ancient Wisdom	107

Lesson No.18

Longevity	108
He Lived 256 Years	115
Body Never More Than Seven Years Old	116
Fruit and Longevity	117
Doctors Do Not Live Long	118
Live 200 to 300 Years	119

Lesson No.19

Water Causes Aging	121
Less Minerals Needed After Maturity	122
Causes of Sclerosis	122
Importance of Water	123
"Mineral Water" Not Beneficial	124
Lime Deposits Cause Stiff Joints	125
Rain Water and Distilled Water Are Safe	126

Lesson No.20

The Wonderful Orange by Dr. Leon A. Wilcox	128
King of Fruits	129
Finest Distilled Water	129
Six Months on Orange Juice by John W. Marshall	130

Lesson No.21

Breath of Life	136
Early Theories of Respiration	136
Physico-Chemical Theory of Respiration Rediscovered	137
Breathing Primary Function	139
Why Man Dies	140
Why Man Lives	140

Lesson No.22

Spiritual Organs	143
Materialism Is A Superstition	144
Recovery From Illness Only Partial	145
The Kingdom of God Within	146
Uncanny Powers of Indians	147

Lesson No.23
- Spiritual Powers ... 150
- Ancient Science of Man ... 151
- Spiritual Intelligence ... 151
- Man's Intelligence ... 153
- Intelligence of Animals ... 153
- Man A Miniature Universe ... 155
- Man Is Dead As He Lives ... 155
- Man Lives in the Spiritual World ... 156
- Parthenogenesis (Virgin Birth) ... 157
- The Kingdom of God ... 158
- Feeble Minds ... 160
- Ancient Science vs. Modern Nonsense ... 161

Lesson No.24
- Physical Purification ... 163
- Function of Breathing ... 164
- Shower of Red Mist ... 165
- Blood Poison ... 166
- Breath Culture ... 167
- Vital Function ... 168
- The Residual Air ... 169
- Gases and Acids ... 169
- The Skin ... 171
- Exhalation ... 171
- Atmospheric Pressure ... 174
- 500 Felled by Gas ... 174
- Suffocation ... 175
- Cold Facts ... 175

Lesson No.25
- Breath of Death ... 177
- Poisons in the Blood ... 178
- Deadly Carbon Dioxide ... 179
- Exhaled Air Is Poisonous ... 180
- Poisons Entering the Body ... 181
- Big Battle for Health ... 182
- Millions Chronically Ill ... 183

Lesson No.26
- Poisoned Air ... 184
- Carbon Monoxide Gas ... 185

Dangers of Poisoned Air Unknown ... 186
Brain Poison ... 187
Pernicious Anemia ... 188
Cerebral Hemorrhage ... 189
Lung Cancer ... 190
Nerve Cells Destroyed ... 191
Black Lungs ... 191
Causes Cancer ... 194
Eat Up the Body ... 194
Smoke, Soot, Acid, Gas ... 195
Tobacco Smoke ... 197
Coronary Thrombosis ... 198
Science Editor Dies of "Heart Attack" ... 199
Cigarette Consumption ... 200
King George VI ... 200
Change Your World ... 201

Lesson No.27
The Common Cold ... 203
Degenerative Process ... 203
Polluted Air ... 205
Acute Ailments ... 207
Air Cells Burst ... 208
Vital Adjustment ... 209
No Complete Recovery ... 210
Hardened Mucus ... 211
An Invisible Foe ... 212
The Aging Process ... 213
The Blood ... 213
Wonders of the Air ... 215

Lesson No.28
Cosmic Air Purifies ... 217
The Home ... 219
Where to Live and Sleep ... 219
Ozone ... 220
Ionized Air ... 222
Shallow and Deep Breathing ... 223
Breathe More — Eat Less ... 225
Air is Life ... 226
Eternal Physical Life ... 227

The New Age	227
Lesson No.29	
Degeneracy of Civilized Man	232
Civilization	234
Alexis Carrel	235
Lesson No.30	
Mysteries of Life	238
Modern Science Is Young	239
Life	240
Science Changes With the Wind	241
Atheists and Evolutionists	242
The Physical	242
The Cosmic Circle	243
The Physical Plane	245
Four Cosmic Bodies	246
Rise Above the Material	247
Physics and Chemistry	247
Intelligence	248
Fourth Dimension	249
So-Called Law of Gravitation	251
End of Creation Theory	252
Nature	253
Two Kinds of Facts	255
See the Invisible From Within	256
Science of Man	257
Ancient Wisdom Destroyed	257
Ancient Science	259
Spiritual Kingdom Within	260
The Ancient Voice	261
The Glorious Resurrection	262
Lesson No.31	
The Sacred Science	263
Dangerous Invention	263
More Pious Fraud	264
Spiritual Life	264
Regeneration	265
Man's Decreasing Powers	266
The Red Dragon	267
Two Laws of Generation	268

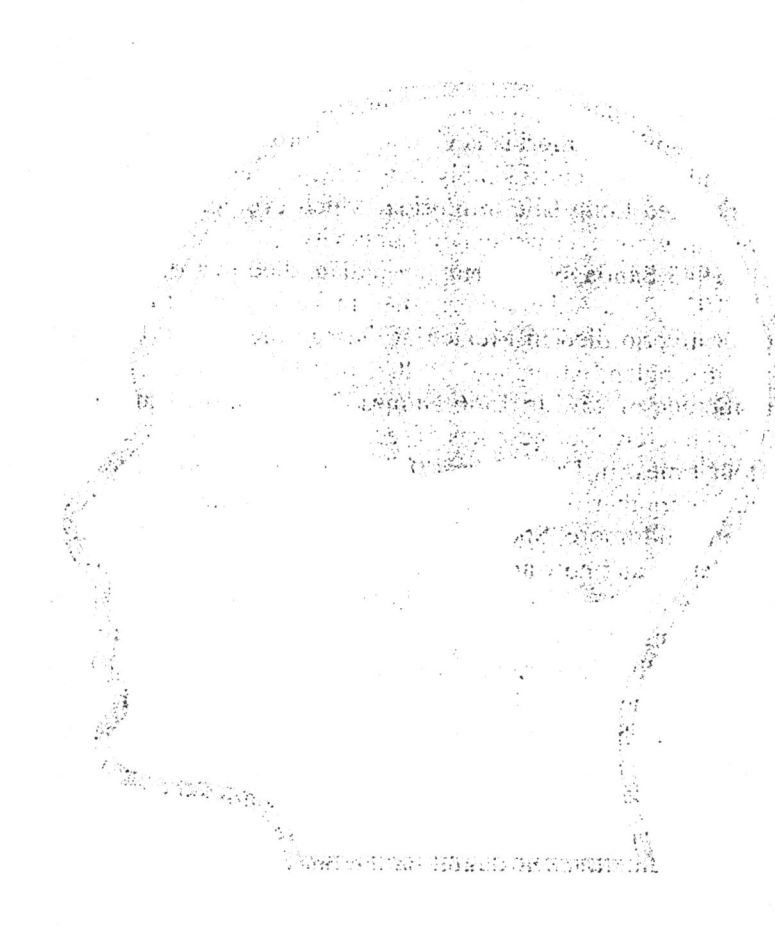

Generation and Death 268
Seven Spiritual Centers 270

Prologue

Life is Creation's greatest treasure for Man in the flesh, and most men should enjoy it much longer than they do. This can easily be done by learning the body's simple requirements, and living in harmony with that knowledge, which is contained in Hotema's late work titled Long Life in Florida. Some men are now living 120, 150 and 200 years and even longer, and what is possible for one man is possible for millions more. Charlie Smith of Florida is vital and vigorous at the age of 119 and says he intends to live considerably longer. Much about him is contained in our work titled Long Life in Florida, which everyone should read, as it's the greatest work ever written on Longevity.

In 1943 Santiago Serviette, an Indian, died in Arizona at the age of 135. In 1936 Zora Agha, a Turk, died in Turkey at the age of 162. In 1921 Jose Calverio died in Mexico at the age of 185. In 1795 Thomas Carn died in England at the age of 207. In 1933 Li Chung-Tun died in China at the age of 256. In 1566 Numes De Cugna died in India at the age of 370. He grew four new sets of teeth and his hair turned from black to gray four times. In December, 1888, when Mrs. Fred Miller was two years old, her mother found a little turtle nearly frozen in an alley near their home in Baltimore. She took the turtle home, named it Pete, nursed it back to health, and now at an age estimated to be more than 100 years, Pete is still the family pet and shows little signs of his age.

All living creatures are ruled by the same laws of Creation, but they do not all live in harmony with those laws. Those that reach the closest to it are those that live the longest in proportion to the length of time required for them to reach maturity of physical development. There are many reasons, most of them preventable, why people die young and hospitals are filled with the sick, whereas others are seldom ill and live three to ten times longer. It would logically seem that living creatures with the higher intelligence should be the ones to live the nearest to the requirements of the laws of Creation; but in action it seems to work the other way, the more intelligent creatures being those who appear to stray the farthest from the straight and narrow path which leads unto life (Mat. 7:13, 14). The large majority of the so-called health writers are not noted for longevity. They seem to die as early as those who read their writings. And some who live 120 years can't tell why they lived so long, as in the case of Diamond, who lived 120 years, wrote a book on long life in 1904 when he was 108, on page 43 of which he itemized "My Daily Menu", which is rather good, yet not one we would recommend.

To reduce longevity to a scientific basis means that we must learn the requirements of the body and supply them. The facts show that living is breathing. We can't die as long as we can breathe — we stop living when we stop breathing. Breathing is the primary function of the body. We can live for weeks without eating, and for days without drinking, but when we stop breathing for a few minutes we stop living.

And right here is the most neglected spot in the entire health field. Why is that so? First, ignorance: second, the claim of science that man lives on what he eats; and third, no one has yet found a way to commercialize air end breathing. This is the discouraging condition we found sixty years ago, when we set out to learn how to live healthy and long. And so, we began almost alone to learn something about breathing and the Breath of Life. The first valuable hint came when we found this: "If we maintain our blood in normal condition and circulation, sickness would be almost impossible. The blood is the life of the flesh. We are what we are by the influence of our blood flowing through our body" (Bernarr Macfadden, in Vitality Supreme, 1910).

Then there came another surprise when we discovered that blood is made of gas. The gases of the air constitute the total composition of the blood. We know that water is the product of the uniting of hydrogen and oxygen gases. When we drink water, we drink gases in fluid form. Blood is gases in fluid form. Everything can be transformed to gas by heat. The earth itself is constituted of condensed gas. This is the source and origin of all things. We have heard of fire damp, ignis fatuus, and will-o'-the-wisp. That fiery element is the Living Gas in what we eat and drink. That Living Gas is all the body uses of what we eat. Of that Living Gas the blood is made. This is the first valuable lesson in dietetics. Scientists talk learnedly and foolishly about protein, carbohydrates, nucleic acids, fats, lipids, etc., ignoring the fact that the ox, elephant, horse and moose live in good health all their days on grass and green leaves.

The next lessons in dietetics is not to heat food and drive out of it the precious, volatile gases which the body uses in its laboratory to make its blood and the products it needs, which includes all the elements mentioned above. Remember this one: Creation never uses second-hand material in its building work. The protein in the food we eat never becomes the protein of our body. That protein has served its purpose, is a used product, and is never used again by Creation in its constructive work.

The living gas in what we eat is all the body uses. The rest is useless waste, and cast off by the body as feces. Hence, most of what man eats goes down the sewer. As gases are all the body uses in making blood and bone and building flesh, consider the condition of the blood, bone and flesh that are made of the poisonous gases that saturate the air of Modern civilization, where health is the exception instead of the rule. If a chemist analyzed the air we breathe and gave us his report, we would be shocked to learn the great amount of poison the body must endure to live in our polluted environment. This subject is so broad and vital, it would take a large book to discuss it adequately. But enough has been said to make a thinking person be more careful about the kind of air he breathes, the condition of the air in his home, and especially in his bedroom, where lack of activity during the night allows the air to stagnate and grow extra foul. That's another reason why people die in their sleep. The polluted, stagnant air in their bedroom paralyzes the breathing center of the brain, and they just stop breathing.

Stagnant air gets foul, like water in a stagnant pool. Keep the air in circulation in homes and bedrooms. Use electric fans for that purpose. Fewer people would die in their sleep if they had an electric fan in operation in their bedroom. This writer is only fifteen years under the century mark, which he expects to pass by many years, as he feels as fit as he did forty years ago. And he is telling the world here, in this work of his, some of the secrets of how he has done it.

To an intelligent, unprejudiced person who can and does think, the information contained in this work may seem simple. But it is the fundamental simplicities that are always difficult to accept, because they are so very simple, and, therefore, unbelievable on that account.

Prof. Hilton Hotema
Tequisquiapan, Qro.,
Mexico.
Alt. 6000.

Why We Live & Why We Die

Scientific investigation shows that —

*We breathe to live and we breathe to die,
We drink to live and we drink to die,
We eat to live and we eat to die.*

In this work it is sought to show how to take DEATH out of Breathing, Drinking and Eating, so man may live 250 to 300 years. As Dr. Robert McCarrison, of the British Army Medical Service in India, reported of a colony of people he found living in a certain Himalayan region, who were active and vigorous, yet so old in years that he was astounded and could not believe their records were correct. But he found no error in their way of keeping time. He said: *"Men well over 200 years of age working in the fields with much younger men, doing as much work, and looking so much like the younger men, that I was unable to distinguish the old from the young."* - K. L. Coe in Correct Eating &Strength, March, 1931.

Introduction

The living cell is animatized by Cosmic Force and Intelligentized by Cosmic Consciousness. The human body is a mass of trillions of cells, guided by Infinite Intelligence. Each cell is a mass of millions of atoms, each of which is a miniature solar system, with "planets" in the form of tiny electrons whirling at tremendous speed around a common center of attraction. The solar system has no use for food. The electrons do not eat. The atoms do not eat. The cells do not eat. Why should man eat?

The Parent Cell begins the body of man and expresses the uncanny intelligence which it inherits from the Cosmic Principle of Creation. This cell does not come from food, and does not depend on food. The body is not the product of food, and should not depend on food. Food cannot sustain what it cannot produce. As the body is not the product of food, it is actually not sustained by food. Infinite Intelligence works in the cell, from the Parent Cell on through the entire existence of the body. Being directed by infinite Intelligence, the cells know spontaneously the functions they are to perform in the construction and maintenance of the body. The innate intelligence of the part the Cell must play in the whole body is a mode of being of all the elements of the body. These elements know their work, and can receive no aid from human hands.

The body cells are seen to understand numerology, geometry, physiology and biology, and act concertedly for the good of the whole. The spontaneous tendency of the cells toward the formation of the organ in the body is a primary datum of observation.

The body, from the Parent Cell on, is built by techniques and directed by Intelligence entirely foreign to the best scientific minds. From the Parent Cell composed of invisible atoms, and out of the invisible Essence of the Universe in the Infinite Air which contains ALL in itself, the body and its organs come into being by the work a cells endowed with Infinite Intelligence. The cells prove by their work that they possess a prevision of the future structure and its purpose, and they synthesize from the atomic substance that appears to be contained in the plasma, not only the building material, but also the builders.

The trillions of cells forming the body are tiny suns and stars, composed of the same cosmic essence and governed by the same cosmic law. A droplet of water forms a tiny microcosm containing a great variety of cosmic chemical elements. The body cell is practically a duplicate of a water droplet, but raised to the high plane of divine animation.

If the cells depend on food and drink, man should continue to grow and live as long as he had sufficient food and drink. According to the press of May 27th, 1937, Srimati Bala, of Bankura, India, age 88 had taken neither food nor water since she was 12. The account said "She is always gay and looks like a child." What is possible in one case is possible in millions of other cases. Abbe N. De Montfaucon De Villars stated in his book that the Ancient Masters ate food only for pleasure and never of necessity (Comte De Gabalis).

Man Eats To Die

The world is flooded with books on food and feeding. No one seems to realize that eating is not natural, but an acquired habit, like smoking and drinking, and that Air is the Cosmic Reservoir of all things, including the substance that builds and sustains the human body. Science shows that the body is built of cells, which are composed of molecules, which are composed of atoms, which are composed of electrons, which are nothing more than whirling centers of force in the ether.

Electrons do not eat, atoms do not eat, molecules do not eat, cells do not eat, and the body is built of and sustained by the cells, and not what man eats. More proof that eating is not only a habit but a bad one, appears in the fact that a sick man often begins at once to recover when given no food, and even shows signs of growing younger.

This could not be, and it would be dangerous for one to fast, if eating was natural and food was needed to sustain the body. Why does man seem to starve to death when deprived of food? That riddle "MANS MIRACULOUS UNUSED POWERS" considers. For more than half a century, the author read books on food and feeding, and closely followed the arguments and explanations. He found those who favored Vegetarianism omitted all the bad features, and the same course was pursued by those who favored Carnivorism. Books favoring Vegetarianism say nothing of the damaging qualities of vegetables and Cereals. Those favoring Carnivorism carefully omit the damaging properties of flesh. These authors lead their readers astray with half truths. A half truth is more dangerous than a lie, as it is more misleading.

The author describes the damaging qualities of all foods, and favors none. What he shows will shock the reader and show him why that group of eminent doctors in the 19th century, after studying the food question from every angle, concluded with this astounding statement:

"We eat to live, and we eat to die."

Why is it possible that we eat to live and eat to die? If we eat to live, how can we eat to die? If we eat to die, how can we eat to live? In this work these puzzling questions are considered and answered.

Lesson No. 1
Physical Perfection

Physical Perfection would be Physical Immortality; and Herbert Spencer formulated the Law of Physical Immortality as follows to wit: "Perfect correspondence would be perfect life. Were there no changes in the Environment but such as the organism had adapted changes to meet and were it never to fail in the efficiency with which it met them, there would be Eternal Existence and Eternal Knowledge" (First Principles).

Breatharianism in Physical Perfection: Man came into physical existence a Perfect Breatharian. God breathed into his nostrils the Breath of Life, and man became a living entity (Gen. 2:7). Nothing was lacking, and nothing more was needed. The Breath of Life supplied all the requirements of animation. The Breatharian needs air only, and nothing more, to sustain his body. In that state of Physical Perfection man has no other wants. The less man needs the more he becomes like gods, who use nothing and are immortal, said the Ancient Masters.

Poverty, Want and Sickness are the work of man. They are the products of his habits which correspond with his desires. He increases his burdens as he multiplies his wants. The less man needs the more complete he becomes. He gains Perfection as he gains freedom from all Wants. The more Wants he has, the less complete he is, and the farther he inclines from Perfection. He changes his world as he changes himself.

Breath of Life

"Every living thing must breathe the air in order to live. The tree breathes the air through its leaves. The leaves are in this sense the lungs of the tree. Insects breathe the air through tiny openings in their bodies. Frogs breathe the air partly through the skin. Fishes breathe the air by taking oxygen out of the water as it passes over their gills. Man breathes the air through the air cells of the lungs." Frederick M. Rossiter, B.S. M.D., L.C.R.P., London. "Hardly anyone understands this science (of breathing), which all should know and practice. It is the air that renews our blood and brings life to all our organs. It is the air that helps to give us balance and to keep our physical and psychic functions in good order . . . Most people use only a third, a quarter or even a fifth part of the lung's total surface." - Professor Edmond Szekely.

Modern Man is the degenerate descendant of the Breatharian. During the millions of years that man has inhabited the earth, his environment and the many habits he has formed have forced his body to adapt itself, under the law of Vital Adjustment, to many evil conditions and harmful substances in order to survive, all of which are foreign to the body and injurious in character.

Stages of Degeneration

Every facet of living existence, the Law of Vital Adjustment, the evidence contained in ancient scriptures, all prove without an exception that Modern man is the product of descending Evolution — Devolution. The scanty evidence that has survived from the remote past, shows that Modern man has slowly descended through the five stages:
1. Breatharianism 2. Liquidarianism 3. Fruitarianism 4. Vegetarianism
5. Carnivorism.

Breatharian indicates a plant or animal that neither eats nor drinks and subsists entirely on the substances contained in the air. It is surprising to know how many such plants and animals there are. The Spanish moss hanging on the trees of Florida gets all its substance from the air, and grows very fast. The big cactus plants on the dry, barren deserts of the southwestern part of the U.S.A. get their substance from the air, their roots serving as anchors to hold the plants in place and complete the Life Circuit.

Magnetism

In our work, "THE NUTRITIONAL MYTH," we stated that no chemist can find in the grand in which grows a giant tree, the ash, mineral, carbon, wood and chlorophyll contained in the body of the tree and its leaves. Nor does a tree consume the soil in which it stands. If so, as evidence of such consumption a depression should surround every tree. A large tree weighs tons, and all the earth still surrounds it that was there when it was a small sprout. The roots appear to serve only as anchors to hold the tree in place and complete the electro-magnetic Circuit of Life.

If corn or wheat is planted in moist sawdust around a magnet or loadstone, growth is promoted without adding more elements. No real relation appears to exist between the elements in the soil and the plant. Examination shows that plants contain elements which never exist in the soil. They are supplied by cosmic rays.

It is the general belief that crops consume the soil in which they grow. Jean Van Helmont did not believe it. In the 17th century he weighed the soil he put in a tub, then planted a tree in it. At the end of four years the tree was six feet high and weighed many pounds. But the soil in the tub weighed the same. The growth of the tree resulted from the elements supplied by cosmic rays. There must be ground connections in the base of plants to complete the Life Circuit. Ground connections are necessary to complete the circuit of electric instruments.

Trees grow not in certain areas, nor grow well in what is termed poor soil. They improve if given proper fertilizer. To make the circuit effective, the right minerals must be in the ground, and there must also be moisture. It is not the tree that needs the minerals. They are required to supply the conditions necessary for proper electro-magnetic action. You know more about motor car batteries. When the battery weakens and runs down, it is recharged. The recharging process does not infuse electricity into the battery. It only changes the chemistry of the battery fluid.

All hibernating animals are complete breatharians during the weeks they sleep in winter. A fasting man who takes only water during his fast has simply added liquidarianism to breatharianism. That is the path back to Physical Perfection. In our work "THE NUTRITIONAL MYTH" we referred to cases of those who are reported to live without eating — Mrs. Martha Nasch who had eaten nothing for seven years; Teresa Neumann, a German mystic, who had taken no food nor liquid for forty years.

The press of May 3rd, 1936, mentioned the cases of Srimati Giri Bala Devi of Patrasayar in Bankura, India, the sister of Babu Lambadar Dey, pleader, as a woman who had taken neither food nor water for 58 years. The account said: "She takes nothing, not even a drop of water. She is always gay and looks like a child. She does not pass stool or urine, and does her house work like any other woman." This woman was reported as being 68 years old, yet she "looks like is child." This appears to be getting close to the Law of Perpetual Youth and Eternal Physical Existence.

One physiologist states that of what man eats, the only parts that do and can enter his blood are the gaseous and fluidous elements. All the rest is nothing but waste that passes through the bowels as feces, playing its part in overworking and weakening the stomach and bowels, with constipation afflicting practically the whole nation. There are very few folks who are free of stomach and bowel troubles, and the markets are flooded with worthless and harmful remedies for these disorders.

Spiritual Potentiality to Physical Actuality

Scholars hold that in Physical Perfection man was entirely free of all wants and desires. The needs and requirements of his body were fully supplied by the Pure Air of His Perfect Environment.

That perfect state was a condition precedent to his coming into physical being. It was the cosmic perfection of his Environment that made possible his evolution from Spiritual Potentiality to Physical Actuality.

It is an established fact that something cannot come from nothing. Men could not become a physical entity had he not existed as a spiritual potentiality. There must be a precedent for every subsequent, a cause for every effect. The cause, whether first, last or anywhere along the chain of causes, must be the comprehensive equal of the effect.

The stream cannot rise above its source. If the rivulet can flow but an inch higher than the sufficiency of its cause, there is no reason why it should not climb the mountain-top and "increase by the force of its own intensity," as is said of disease. No effect can produce its own cause. The Universe could not create itself, which is equivalent to the admission that no part thereof could create itself.

Universal existence is eternal existence. What appears as the physical world which we call Nature, is the materialization of Spiritual Potentialities. Had man not existed Spiritually, he had never become a physical reality; and his appearance as a physical entity is conclusive proof of the perfection of his Environment at the time of his Physical beginning. That Environment had to and did possess the perfect powers and requirements that evolved man, not from an ape, but from Spiritual Potentiality to Physical Actuality. It fulfilled every need, every demand, everything, it had been fatal.

The Way To Improve Physical Man

Advanced scholars point out that there could have been in man's physical beginning no unfilled wants; otherwise physical being had been impossible. They show that the only way to improve physical man, is to reduce his wants and decrease his economic burden.

But physical science is not interested in any course that raises man to a higher plane. For nothing must be done, nothing must be allowed to happen, that will disturb or derange the fixed social order of civilization.

This order is the product of ages of planning and scheming. It depends upon and is sustained by man's wants and desires, and the constant effort made to promote and increase them. To that end all education is directed. Every branch and department of civilization leads away from Perfection. The movement away from Perfection begins with the child in school and continues all through life.

So Called Civilized View

On January 12th, 1951, Frank W. Abrams, Chairman of the Board, Standard Oil Company of New Jersey, made an address before the National Citizens Commission for Public Schools, and his address was published and widely circulated. Among other things he said: "There can be no doubt that we are talking about something very fundamental to business when we talk about education.... If only to maintain and expand its markets, the business world has at least as big a stake as anyone in the achievement of an educated, productive and tolerant society.... There is definite correlation between education and the consumption of commodities. Education has done more to create markets for business than any other force in America."

This is the orthodox view, and according to that view, the purpose of education is to maintain and expand the markets of business, and to create a demand for commodities. To that end billions of dollars are expended annually in the education of the children of America.

The constant cry of Commercialism is to consume more, create new markets and new demands, promote the production of commodities, employ more wage slaves, and increase the economic burden.

Why The Truth Is Suppressed

The art of scientific living is so lightly regarded that it receives no attention. He who is so far ahead of the multitude as to oppose the social pattern is promptly silenced, disgraced and liquidated; and the press carries large headlines proclaiming that an enemy of social progress has been found and jailed. The deceived multitude believes.

One's teaching may be in harmony with God's Plan of Life, the Law of Perfection, and the Science of Cosmic Economy. But that teaching does not harmonize with civilization's artificial world, or support its social pattern; therefore, it cannot be accepted and supported by any institution or any form of government. It must be suppressed *"for the good of the people."* Tell us how long man's created wants and unnatural desires will mean money for Commercialism, and we will tell you how long man will remain in his present condition of degeneracy and economic slavery.

Economic Freedom

It is interesting and important to note that as man moves back toward primal Perfection, his wants decline and his economic burdens decrease. We thus learn what these burdens are and whence they came. We see them as the product of man's created wants and unnatural desires which Perfect Man had not. Man has produced them, and he can destroy them. It was not until man began to form habits and adopt practices which created wants that he began to decline and degenerate. He was deceived then, as he is now, by the illusion of progress as he developed new habits that increased his wants. He considered then, as he does now, that each new invention was a mark of progress while he and the doctors were puzzled by the fact that his health continued to decline and his life-span to decrease. Economic freedom is the first step back toward man's high estate of primal Perfection. Every animal, in its native state, has economic freedom. Man is the only economic slave on earth. He has made himself such by his unnatural wants and acquired desires.

In complete freedom from every want, to be dependent upon nothing, man's mind and senses are under control. He is released from the consequences of action, which are bonds and chains, binding down those who are the slaves of want and desire.

Lesson No. 2
The Living Cell

"Life is the expression of a series of chemical changes." - Osler in Modern Medicine P. 39. Modern science has spent years trying to define Life, and some of the many definitions advanced appear under "Life" in our work titled "SCIENTIFIC LIVING."

In "The Book of Popular Science" it is said: "What is Life? That is another question and the answer is one of the profound mysteries. Science has explored life down to a single cell of living matter, but exactly what makes the cell alive is not known."

One of the main concepts of science, the "living protoplasm" which has been regarded by science for a century as the source of life, was exploded in 1937 by group of America's foremost scientists, who showed that protoplasm is composed of numerous, ordinary particles, visible by modern methods, and not one of these particles are alive. Protoplasm is the sticky, whitish substance of which living tissues seem to be made. It was so named in 1840 from the Greek Protos (first), and plasm, a thing formed, and means "the first citation."

Dr. E. N. Harvey, of Princeton University, related how, under the microscope, protoplasm has literally fallen apart in the whirling centrifuge and fine testing methods. It appeared to be composed of cells, bits of fat, granules of colored matter, proteins, threads, hollow bubbles, nuclei, minerals and a complex of other substances, — "not one of which," he said, "can be considered as living except insofar as it is indispensable for the continuance of Life." Thus, the mystery of Life remains unsolved so far as Modern science is concerned. Osler declared that Life is neither an entity nor a principle, but only the expression of a series of chemical changes. Then came the famous Carrel whose experiments showed that Osler is wrong. Carrel said, "The childish physico-chemical conceptions of the human being, in which so many physiologists and physicians believe, have to be definitely abandoned" (Man, The Unknown, P. 108).

This leaves Medical Art with no law of Physiology, and it cannot have one until it knows what makes the body function.

Chiropractic Law Of Physiology

The Science of Chiropractic discovered the law of Physiology, and the law was formulated by the great Willard Carver, who died in 1943 at the age of 80. He founded the Carver Chiropractic College at Oklahoma City in 1905. Carver wrote: "Organisms that move in conduct we term animation, are not alive. Life is imminent in such structures, but not inherent in them. Life flows through them. We speak of a wire as being alive whilst a current of electricity is flowing through it. If we cut off the current, the wire becomes dead; yet the wire itself has assumed no visible change. "Life, as it is observed in this material existence, is the action of Matter under the operation of Forces. The body of man is alive in the sense that it is animated by the Life Principle flowing through it." - (Psycho-Bio-Physiology).

According to Carver, the Living Organism is an instrument and its animating power is the Life Principle. But this old doctrine of the Ancient Master, who held that the Cosmic Spirit animates the body of man (Jn.6:63), "is a mere superstition," according to Modern science, and accepted now only by the layman and the Chiropractors. No satisfactory theory of the evolution of man, as advanced by the evolutionist, can be sustained so long as the Genesis of Life upon this planet is shrouded in darkness. The basic factors and causes of evolution are bound up in the question of Life itself.

The evolutionist begins with the Living Cell. Once given a primordial life all cells are capable of reproduction, and modern science constructs man physically, vitally, intellectually and morally. According to Modern science, all these properties appear in matter as the result of "the expression of a series of chemical changes." Preposterous: that exceeds the work of the magician in pulling white rabbits out of a hat.

Where Did The Living Cell Come From

The evolutionist makes no attempt to explain the original appearance of the living cell itself. Neither does he explain the nature nor cause that originated it. He does not explain the original division of living organisms into male and female. He does not explain the phenomenon of intelligence displayed in the conduct of all living things.

Prof. Hotema   Man's Higher Consciousness   Lesson No.2

Modern science explains none of these things. On the other hand, it most unscientifically relegates them to the realm of the "Unknowable."

Modern science confesses itself baffled at every point when it would explain how life evolves from non-life, how sensation evolves from non-sensation, how intelligence appears in living creatures. It fails to explain these phenomena just as it fails to explain how intuitive intelligence rises into rational intelligence, or how unmoral perceptions rise into moral conceptions. The vigilant biologist traces life to the nucleated cell. Here, in the department of Protozoa, he becomes bewildered. He misses the connecting link. He fails to discover that subtle principle which enters into and converts inanimate substance into living organism. This lack of basic knowledge caused Carrel to state that, "Our ignorance (of the body and its functions) is profound.... The mechanistic physiologists of the 19th century... have committed an error in endeavoring to reduce man entirely to physical chemistry (Man, The Unknown, P. 4, 34). In more definite terms, modern physiologists of the orthodox school are still in darkness as to the body's requirements and functions. What the body needs to sustain it and why it functions are things unknown to them. The field is wide open for the consideration of Man, with almost nothing definitely known by physical science, and any reasonable theory advanced may be as correct as any other.

Is Eating Necessary

"Appetite comes with eating." - F. A. Ridley.

In our work titled "THE NUTRITIONAL MYTH," we stated that it is more difficult to explain why man should eat than to show that he should never eat. It is said by some unorthodox physiologists that man should not eat; that what he does eat does not sustain his body; that the cells of which his body is composed are self-existent and self-sustaining, as are all other objects that are composed of atoms as the body cells are; and that man should have no more need for food than a stone or a star, since these also are constituted of atoms the same as are the body cells.

Science knows that the Parent Cell begins the body and builds it by a process of cell division and subdivision. Food does not build the cell or the body. Neither can food sustain the cell nor the body.

New cells are not produced by food but by the division of pre-existent cells. The new cells replace the disintegrated cells; modern science knows that. It knows that the new cells are not the product of food. Yet it insists that the body is nourished and sustained by what man eats. Consistency of thought demands that we proceed in our processes in a direct line through infinite time to infinite results. If food does not produce cells, if all cells of the body are produced by the process of division and sub-division of pre-existent cells, then food does not and cannot nourish and sustain the cells. In that case the cells need no nourishment, cannot receive or use it, and are self-existent and eternal.

If food can produce and sustain Living Cells, if unintelligent matter can produce intelligence, there is no reason why man should not progress to infinite capacity by virtue of the power residing in the food he eats.

Elimination

Experiments consistently show without an exception, that elimination is much more important than feeding. Many refuse to recognize that fact. The maintenance of the vital condition of the body is more intimately and immediately related to and dependent upon the excreting part of the physiological process than it is upon the supply of new aliment. Many refuse to recognize the fact that feeding may be suspended for a considerable period without causing anything more serious than loss of weight and strength. The records show that people have lived without food all the way from forty days to forty years, and still lived in good health. But the elimination of the effete substance produced by tissue disintegration cannot be checked for even a few minutes, in warm blooded animals, without inducing fatal results.

Every act of respiration is in effect the leading process of excretion, and the only process of animation. For to stop the breathing is to stop the living (LONGEVITY — Chapter 3). The products of cell disintegration are liquids and gases, of which the gases from the greater part. The major portion of the gases is eliminated via the lungs. The rest leaves the body through the kidneys and skin. The liquids also leave the body via the kidneys, lungs and skin.

Cells Are Not Produced By Food

We reiterate that what man eats and drinks does not produce cells or nourish them. The cells all come from the Parent Cell, the beginning point of the body. The body cells do not depend on what man eats and drinks. They are not nourished — that is another illusion. The cells are far above the nutritional level, are independent of nutrition, and are self-existent and eternal — a fact, demonstrated by Carrel.

Upon the demise of the body, the tissues, glands and organs disintegrate, setting free the cells that constitute these structures; and the immortal cells return to the Cosmic Ether, whence they Came. The cells are perpetual and indestructible, as are all primal elements of the Universe, to which order the body-cells belong.

Early Men Were Breatharians

These facts have constrained unorthodox physiologists to advance the theory that the first men in the early days were Breatharians, deriving from the Divine Breath of Life (Gen. 2:7) all the substance they required for its existence, and they lived from eighty thousand to a hundred thousand years; or until they were ready to return to their spiritual home, as stated in our work on LONGEVITY, Chapter 12.

The human body is now known to be a composition of trillions of cells. Each cell is a mass of millions of atoms, each of which is a globular system with "planets" whirling with tremendous speed. What use, asks the unorthodox physiologist, has such a system for food and drink? What use has our solar system for food and drink?

The Miraculous Cell

The body cells are endowed with unsuspected powers and astounding properties. Despite its smallness, the cell is an exceedingly complex organism that does not in the least resemble the favorite abstraction of the modern chemist — *a drop of gelatin surrounded by a semi-permeable membrane.* While the structural complexity of the living cell is disconcerting to the orthodox biologist, its chemical constitution is still more intricate. A droplet of water forms a tiny microcosm containing, in a state of extreme dilution a great variety of cosmic elements. The body cell is practically a duplicate thereof, but raised to the exalted plane of divine animation. Carrel says that the Parent Cell and the Cells that come there from are "not made of extraneous material, like a horse" (P. 107).

He means that the cells are not composed of what man eats and drinks. He states that while the body is composed of cells as a house is of brick, the body is born of a cell that makes more cells, just as though a house originated from one brick, "a magic brick that would set about manufacturing other bricks. Those bricks, without waiting for the architect's drawings or the coming of the bricklayers," says Carrel, "would assemble themselves and form a house" and all parts thereof.

Lesson No. 3
Physical Immortality

"The human frame as a machine is perfect. It contains within itself no marks by which we can possibly predict its decay. It is apparently intended to go on forever." - Monroe. John Gardner, M.D., in his work "Longevity," wrote: "Before the Flood men are said to have lived 500 and even 900 years. As a physiologist I can assert positively that there is no fact reached by science to contradict or render this as improbable. It is more difficult, on scientific grounds, to explain why man dies at all, than it is to believe in the duration of human life for a thousand years."

Dr. Foissac wrote in his book on "Longevity": "The long life of the biblical patriarchs is a fact more rational and more in accord with the known laws of physiology, than is the brief existence of the men who inhabit the earth today (Part II, Chapter 22). After years of investigation, the famous Metchnikoff declared that deterioration of the bodily structure and old age are due to minute quantities of poisonous substances in the blood.

Why Man Degenerates

In his book, "Prolongation of Life," Metchnikoff furnishes the first logical explanation in modern times of the degenerative changes occurring in the body, and why. His findings have been confirmed by leading researchers, including such noted doctors as Crile, Emphringham and Carrel. 1. Crile said, "There is no natural death. All deaths from so-called natural causes are merely the end-products of a progressive acid saturation." 2. Emphringham declares, "All creatures automatically poison themselves. Not TIME but these toxic products produce the senile changes that we call old age." 3. Carrel asserted, "The cell is immortal. It is merely the fluid in which it floats that degenerates. Renew this fluid at proper intervals, and give the cell proper nourishment upon which to feed, and, so far as we know, the pulsation of life may go on forever.... Quickly, involuntarily, the thought comes: Why not with man? Why not purge the body of the worn-out fluids, develop a similar technique for renewing them — and so win immortality?" (Man, The Unknown).

Carrel was great, but he missed some vital points. He could not rise above the medical theory of cell nutrition. We show in our work "THE NUTRITIONAL MYTH" that the cells of the body are self-sustaining, self-existent, eternal. They are far above the nourishment level.

In the attempt to nourish cells which need no nourishment, lies one of the errors that hurry man to the grave. Feeding increases the degenerative process, whereas fasting, the opposite course, institutes the opposite effect — the process of rejuvenation. The body is composed of trillions of cells. Carrel shows that all organs and structures that constitute the body as a whole, come from the original Parent Cell by a process of cell division and sub-division. That fact is definitely stated in all text-books on anatomy. It is highly important to note that Carrel further states that the Parent Cell and the Cells that come from it, are "not made of extraneous material like a house." He means that the cells are not made and composed of what man breathes, drinks and eats.

That being an admitted fact, then as the cells are not the product of food, they are independent of food and are not sustained by it. What food does not produce it cannot sustain. As the body is not composed of "extraneous material," we cannot feed it with such material, as such material is foreign to its constitution. The body is composed of cells, not of food; and the cells are composed of molecules, which are composed of atoms, which are composed of electrons, which are whirling centers of vibratory force in the ether. Recent observations show that the entire Material World is only a visible manifestation of varying wave forms. Dr. H. H. Sheldon, University of New York, wrote: "We live in a world of waves. The further we delve into the ultimate structure of Matter, the more obvious it is that nothing exists except in wave form. Electrons, long thought to be the ultimate particles of which all matter is formed, have now been shown to have a reality only as a wave form, while no atom consists of a bundle of such waves."

This means that the cells consist of a bundle of such waves, as cells are composed of nothing but atoms. This further means that the body consists of a bundle of such waves, as the body is composed of nothing but cells. So far as such physical structures as man and animals are concerned, this doctrine might be contested by the old school of scientists; but Sheldon says: "We as individuals undoubtedly have no existence in reality other than as waves, multitudinous and complicated centers, perhaps, in what we call the other.... We are analogous, in a sense, to the sounds that issue from a piano when a chord is struck, or when a symphony orchestra sounds." Electrons do not eat, atoms do not eat, molecules do not eat, cells do not eat, and the body does not eat.

Then why does man eat? That question science cannot answer; but we answer it according to Cosmic Law in our work, THE NUTRITIONAL MYTH." We said: "Cell and body nutrition is a myth. What man consumes does not supply cell nutrition by assimilation as taught by science. The ingested substances merely produce activity in cell function by stimulation and not by nutrition. Two types of stimulation seem essential for the function of living cells: Vital and Chemical. The nerves supply the Vital from the Life Source, and air, food and liquid supply the chemical. The ingested substances contact and stimulate the cells into action, and pass from the body through the eliminative channels, as flowing water turns the wheel of a mill, thus activating the machinery in the mill that does the grinding." - P. 17.

The Transportation System

Consider the water and the water wheel. The water below the wheel is the same that activated the wheel as it passed. The same substance that leaves the body as waste is the same substance that enters the body as air, liquid and food. In their passage through the body, as the water passing the wheel, they stimulate and activate the cells which compose the organs and tissues which constitute the body.

The substances man breathes, drinks and eats enter the body and pass into the blood. The blood, as a transportation system, carries the substances to the cells where they stimulate and activate the cells, but pass on without becoming a part of the cells — noting, the water passes on without becoming a part of the wheel or the mill. And as the water activates the wheel in passing, it activates all the machinery of the mill.

Carrel, as an orthodox medical doctor, went no farther than the cell and held to the medical theory of nutrition. His experiments showed the cell to be immortal. When the fluids fail to carry off what he regarded as the excrement of the cells, he found that the cells degenerate, grow senile, and show signs of dying. Each time they were rejuvenated by a cleaning process. Here again Carrel missed a vital point. Body cells do not die; and he himself declared them to be immortal. The cells simply sink below the life-level of vibratory action because their magnetic poles become corroded by the acids that are not carried off and eliminated by a clogged blood stream. Naturists know that patients recover quickly under fasting because fasting reduces the amount of pollution in the blood; thus, freeing the cells of the injurious effect of the acids. Crile said that "all deaths from so-called natural causes are merely the end-point of a progressive acid saturation." In fasting is the process of renewing the fluids mentioned by Carrel. That is also the process of rejuvenation as many experiments have shown. This startling knowledge leads physiologists to declare that decrepitude and physical death, except by ancient, are due to: (1) polluted blood system and the (2) accumulation of garbage.

This results in a corrosion of the magnetic poles of the cells, making them incompetent to receive the animating amount of vital vibrations of the Life Force, and they fall below the level of animation. Carrel called that condition the death of the cells.

If we keep the cells from dying, we keep the body from dying. For the body is composed of nothing but cells. Neither do the cells die. They are immoral. But when their vibratory rate falls below the life level rate, the body accordingly sinks below the life level plane: and that state is what we term physical death.

Blood Purification

Elimination therefore appears far more important than feeding. If blood purification by the kidneys were stopped, man would die in three to five days. If blood purification by the lungs were stopped, man would die in three to five minutes. In the matter of health and life we always come back to the blood. "It is merely the fluid in which the cell floats that degenerates," said Carrel, and he added, "Why not purge the body of the worn-out fluids, develop, a similar technique for renewing them — and so win immortality?" That simply seems to be the secret of physical immortality. With every poison that man can inhale, drink and eat which does not kill him instantly, his Vital Stream is changed from the River of Health and Life to the Pool of Pollution and Death.

In due time, after much suffering, man dies "of a progressive acid saturation" of his blood, tissues and cells, in fulfillment of the cosmic law that he reaps as he sows (Gal. 6:7).

Lesson No. 4
Vital Adjustment

Carrel and Lakhovsky are the greatest Scientists of the age so far as the constitution of Man is concerned. The former dug deeper into the body and its processes perhaps than anyone else in modern times, while the latter swept the universe in his search for the secret of Life.

Carrel devoted a chapter of his work, Man The Unknown, to the important subject of Adaptive Functions, and declared them to be responsible for the duration of man's life. He showed why man's body, composed of soft, alterable matter, susceptible of disintegration in few hours, lasts longer than if made of steel. He wrote: "Not only does he last, but he ceaselessly overcomes the difficulties and dangers of the outside world. He accommodates himself, much better than the other animals do, to the changing conditions of his environment. Such endurance is due to a very particular mode of activity of his tissues and humors. The body seems to mold itself on events. Instead of wearing out (collapsing) it changes. Our organs always improvise means of meeting every new situation, and these means are such that they tend to give man a maximum duration. The physiological processes... always incline in the direction leading to the longest survival of man. This strange function, this watchful automation with its specific characters, makes human existence possible. It is called Adoption."-191-2.

People know so little concerning the adaptive functions of the body. Hence, it will be difficult for the reader to understand that the weaker his body becomes the longer its duration under adverse conditions.

Carrel himself appeared not to know that man "persists in living despite physical, economic and social upheavals," because of the fact that as the "body seems to mold itself on events," it suffers in the process a corresponding weakness which will be explained as we proceed, and the explanation will be so new to the student that he will miss some vital points unless he reads this lesson several times.

Disease Germs

We shall begin by referring to the fraudulent claim that the world is filled with evil entities which attack healthy persons without reasons and the result is "disease," — something so dangerous that it must be combatted and cured. A deceived world believes in this and rejects evidence that exposes it. Carrel believed in it to the extent that he asserted, "scientific medicine has given to man artificial health" (P. 311).

It requires great prejudice to blind a scholar so completely that he cannot see the obvious absurdity of such a statement. According to cosmic law, man reaps as he sows; and that law has no exception (Gal. 6:7). We shall use that law as a guide to lead us through the wilderness and confusion created by Carrel's "scientific medicine."

Good Health Is Not Immunity

Doctors volubly discuss the theory of "resistance to disease," and Carrel believed in it. Supposedly good health makes man "immune" to disease — breeding influences, to the attack of germs, to vicious habits, to hostile environment. The theory of "scientific medicine" is that man is attacked by disease because of a weakening of the natural body defenses. Alfred Pulford, M.D., M.H.S., F.A.CT.S., a medical practitioner of fifty years, wrote: "The exciting and contributing causes of pneumonia may be, and are, legion; but they all simmer down to the one point, viz, the breaking down of the natural body defenses (Truth Teller, April 1944, P. 2).

In his daily column in the press of June 8th, 1944, Irving S. Cutter, M.D., said: "Now is the time to think of getting rid of that chronic winter cough. Yes, bacteria are responsible. But their very existence with all the irritation they can create, means that basic resistance (of the body) is too depleted to throw them off."

The better doctors know the germ theory of disease is false. If it were true, neither man nor beast could live long. They would be literally devoured by "disease germs." Natural science shows that correspondence must prevail as between the living organism and its environment. That is the positive, primal, and fundamental condition of Existence.

It was impossible for primitive man to come into physical being unless the proper, perfect and harmonious condition of his Environment prepared the way. Furthermore, the health condition of man's body can never be any better than the health condition of his Environment, with the condition of which his body must always be in correspondence.

Bananas grow not in a cold climate because their constitution does not correspond with such climate. Salt water fish does not live in fresh water because their constitution does not correspond with such water — that is natural science. Man is the most perfect of all creatures, and is able to rise superior to bananas and fish by reason of his body's "watchful automatism with its specific characters" which "make human existence possible." He can modify the condition of a hostile climate or adjust himself to it as to live for a limited time in the hottest and coldest regions on earth. The Law of Correspondence in this case means that a condition of harmony must exist as between living things and their environment, or they will die and disappear. That does occur occasionally, causing certain plants and animals to become extinct.

Spencer formulated the Law of Eternal Physical Existence and sought to show that Death was the end result of environmental changes which the living organism had not adapted changes to meet, thus creating a condition of discard and friction that sent the body down to death.

The human body is so perfectly constituted that it possesses the power to prolong its duration by adapting itself to conditions so adverse that they would otherwise cause not only early death, but instant death in some cases.

On this point Dr. Charles W. Greene wrote: "As the air exhaled from the lungs contains a large proportion of carbon dioxide and a small amount of organic matter, it is obvious that if the same air be breathed again and again, the proportion of carbon dioxide and organic matter in it will increase until it becomes decidedly unfit to breathe. It is remarkable fact that the organism, in time, adapts itself to a very vitiated atmosphere, and that a person soon comes to breathe, without sensible inconvenience, an atmosphere which, when he first enters it, feels intolerable. But such an adaptation can occur only at the expense of a depression of all vital functions, which must be injurious if long continued or often repeated" (P. 286). There is a definite statement of what occurs in the body as its "watchful automatism with its specific characters snake human existence possible" in an atmosphere so badly poisoned that it feels intolerable when one first enters it. The adaptation occurs *"at the expense of a depression of all the viral functions, which must be injurious if long-continued or often repeated."*

That is how the body builds up "basic resistance" to inimical influences and unhealthful conditions. We must first weaken the body's vital powers before it will submit without protest to "the dangers of the outside world" and the evil effects of bad habits.

Immunity Reduces Power To Resist

The secret of Vital Adaptation is the riddle that puzzles the doctors. They term it immunity. The body acquires immunity to dangers by reducing its powers to resist them. Exactly the opposite of what medical art teaches. We shall recite another instance of this weakened condition which medical art terms immunity. Greene continues: "This power of adaptation is well illustrated by an experiment of Claude Bernard. He showed that if a bird is placed under a bell-glass of such size that the air contained in it will permit the bird to live for three hours, and the bird is removed at the end of the second hour, when it could have survived another hour, and a fresh, healthy bird is put in its place, the latter will die at once" (Kirks Physiology, revised by Greene, P. 304).

According to medical theory of "vital resistance," the fresh healthy birds should have resisted the effect of the polluted air in the bell-glass and lived for three hours. But it died immediately; whereas, the other bird that had been under the bell-glass for two hours and suffered a certain degree of debilitation by reason of breathing the poisonous exhalations of its own body, could have lived for another hour.

This illustrates what Carrel means when he says, "Our organs always improvise means of meeting every new situation; and these means are such that they tend to give man (living creatures) a maximum duration. The physiological processes (of all animals) always incline in the direction leading to the longest survival of man."

Man in civilization is born into and grows up in an environment of polluted air that may kill in a day a wild Indian of the hills, whose vital body would react so violently to the shock, that death may soon result.

No matter how repugnant or destructive a thing may be, we can endure it, provided time is given to secure the efficient operation of the body's power of adjustment, whereby is prevented a violent swaying of vital activities from one extreme to the other. Only sudden and violent changes become immediately destructive to life, even sometimes when it is a change from evil to good habits. By the reduction of its vitality, the body pays in the process of becoming "immune" to discordant conditions, to poisonous substances, to evil habits.

That is the reason why the vigorous Indians of America became a "dying-race" when they came in contact with the enervating habits of the white man of Europe who survived in spite of the evil effects, because he was born in and grew up under those health destroying conditions. His body was accustomed to them, was adjusted to them; but the body of the healthy Indians was not, and they died like flies.

All evidence proves conclusively that the more vital the organism, the more quickly it succumbs to unhealthful conditions and harmful practices. That is another paradox. We live in a world of illusion.

There is a natural condition of Vital Adjustment to unhealthful conditions and harmful habits, but no Vital Resistance. That is another fallacy that belongs in the same category with the absurd theory of "contagious diseases." It is a paradox that the body, in a weakened condition, will tolerate and endure longer than a more vital body, the various evil practices and inimical influences which it cannot control, and which it must endure or die. Carrel says "the body perceives the remote as well as the near, the future as well as the present," and it prepares accordingly (P. 197). Creative Intelligence knows that the weakened body will endure more and live longer under adverse conditions. This appears paradoxical, yet the truth of it is, every scientist can demonstrate for himself.

Conditions That Destroy Health

The condition of the living organism is governed by habits, environment and climate. If these are good, vigorous health and longevity are the rewards. Conversely, when this trio of factors is bad, the body is forced to adjust itself to endure them, or perish. It is either endurance or death. But for the Law of Vital Adjustment, the race had perished ages ago. As it is, a gradual degenerative process through the ages has reduced man's life-span from eighty thousand years to an average of much less than eighty. It is Vital Adjustment, not Vital Resistance, that enables the body to survive for a few miserable years, with aches and pains, in air so foul that one in vigorous health would be in danger of dropping dead by suddenly coming in contact with it, as the vital bird thrust into the air that was polluted by the exhalations of the previous bird under the bell-glass.

Danger Of Smoking

The Law of Vital Adjustment makes it possible for the body to tolerate man's evil habits. That makes it possible for the smoker to endure and even enjoy his poisonous pipe. The same pipe would make a vigorous non-smoker ill, or might cause death, as death in such cases has been reported. The vital youth in his ignorance takes his first smoke. His vital body reacts with such vigor against the dangerous poison that sickness results. The degree of his sickness is the measure of his vitality. The more vital his body, the sicker he is.

Here are the poisons found in a chemical analysis of tobacco, nicotine, carbon monoxide, carbon dioxide, ammonia, methane, methylamine, hydrogen-sulphide, furfural, nicotelline, pyrrole, pyridine, picoline, lutidine, collidine, formaldehyde, carbolic acid, prussic acid, arsenic. Chemical analysis shows that cigarettes contain the following active poisons: furfural, acrolein, diethylene, glycol, carbon monoxide, pyridine, nicotine, carbolic acid, ammonia, and a host of tarry substances.

The tarry substances in tobacco smoke are approximately 2.25 percent. They stick to the walls of the lungs, forming a coating that obstructs the free passage of oxygen into the blood. Those who must breathe air laden with tobacco smoke are injured almost as much by it as are the smokers. These are the poisons to which the body of the smoker adapts itself in order to survive. It is either endurance or death.

The use of purgatives and laxatives in the case of constipation illustrates the work of the Law of Vital Adjustment. As these poisonous substances are used to force the bowels to move, the body slowly adjusts itself to them and from time to time the dosage must be increased to make the body act, or some other poison more powerful must be used.

As the horses grow weary drawing the heavy load, the whip must be used harder to drive them on. As the body weakens from constipation and the poisons that force bowel movements, the dosage must be stronger or more powerful poisons used to make the body act.

The youthful smoker disregards the warning reaction of his body. He continues the harmful habit, and gradually the prisons of the tobacco weaken the nerves and reduce the vitality to where it can fight back. Now he can smoke in comfort and experience immediately ill effect. He has established "immunity' say some people. But in doing he is taking the short-cut to the grave. The body slowly sinks in process of degeneration and slow suicide under the power of the dea enemy.

We may well say that what is termed the Law of V Adjustment is also the Law of Vital Reduction. It is the process of v reduction that brings the body into subjection to any devitaliz substance, influence, habit or practice.

Toleration By The Body

The body's vitality is reduced to save it from sudden death prolong its duration. Instead of dropping dead, one dies by inches, and the process of slowly dying one suffers until the body can endure more and the grave ends all. Great is the power of the body to adj itself to conditions and poisons that would destroy it; far greater th man can imagine. The Law of Vital Reduction will enable the body adjust itself to the point where the opium addict can take at one tim dose of dope so large that it had killed him quickly had he at first take dose that large.

The law may cause the body to adjust itself to the point wh the venom of the reptile will fall to kill. The Grit of October 14th 19. reported the case of Bill Haast who is said to be in that condition. T account is as follows — "He is full of snake poison himself, for when first began handling the reptiles he set out to make himself immune their bites. The serpentarium owner inoculated himself with larger a larger doses of snake poison and now believes he is the only man in world genuinely immune to all type of snake venom. He has been bitt by nine cobras and many other times by rattlers, mocassins, corals, a various other kinds of killers."

As the Law of Vital Adjustment operates in the case of smoki and other bad habits, so has it operated in the case of eating. When m first began to eat, the substances that entered his then rudiment stomach were poisonous to his body, as in the case of tobacco. persisted in eating and the Law of Vital Adjustment brought his bo into harmony with the habit.

But such on adaptation could occur only at the expense of depression of all the vital functions, which must be injurious if long continued or often repeated. That is the law and there is no exception to its operation. Some substances man eats are still poisonous to him after all the long ages he has followed the practice. His body has never been able to adjust itself completely to some of the things he eats.

Tea and coffee still make sick some who drink these, and no one can avoid their harmful effects. As the flesh eater becomes a vegetarian, the Law of Vital Adjustment goes into operation, and the time comes when even the odor of the meat market or of boiling beef, which was formerly so fragrant to him, becomes obnoxious to him. These are facts that are unbelievable in him who knows them not by actual experience.

Immunity

What medical art terms "immunity" is a condition produced by reducing the body's vitality. The reduction is accomplished by weakening and dulling the nervous system, and it may have far reaching effects.

The Federal Bureau of Census discloses the shocking fact that the number of persons 15 to 19 years old in this country in 1950 was 14 percent less than in 1940, whereas the population in the same period increased 19,500,000. The startling decrease in the number of young people may be the damaging effects of vaccination and inoculation — in an effort to make man disease-proof. Perhaps those who manage to survive the process and live long enough to reach adulthood, will never know the meaning of the term *good health*. The damaging effect of vaccinating and inoculating all school children seems to appear in the young men of draft age. The press of June 21st 1952, reported that 11.6 percent, or 1,443,315 of the 12,416,129 men it classified from 1948 up to April 30th under the draft system have been put to Class 4-F, as unfit for military service.

This includes the states, the District of Columbia and the territories. The Canal Zone, where the medical rule is less rigid in the vaccination and inoculation of children, had the lowest percentage of 4-F's.

Acute diseases are nothing more than symptoms of the body's reaction to internal poisons, or a condition of "acid saturation" as Crile calls it. There is no absolute immunity to internal poisoning. The body cannot be saturated with poisons and not suffer a corresponding degenerative change. The condition of apparent immunity to poison, to bad habits, or to so-called disease, arises because the poisonous substances, or the vaccines and serums, dull the nerves and reduce the body's vitality, making the body unable to react against the poisons that a more vital body would cast off in the eliminative process which are termed diseases. Man gains so-called immunity by paying the price. That price is a dulling of the nerves that reduce the body's vitality, and that is the short-cut to the cemetery.

The wholesale practice of vaccination and inoculation is poisoning the body, reducing its vitality and decreasing its duration so fast that the death-rate after the age of 95 is steadily rising. Professor C. H. Forsyth of Dartmouth College was reported in the press of July 30th 1029, as stating: "The expectation of life from the age of 45 [onwards], is the lowest of which we have any record. Far lower than it was forty years ago — and it is still falling."

Lesson No. 5
Body Changes

"Man ceaselessly overcomes the difficulties and dangers of the outside world. He accommodates himself to the changing conditions of his environment. The body seems to mold itself on events. Instead of wearing out, it changes. Our organs always improvise means of meeting every new situation; and these means are such that they tend to give us a maximum duration. The physiological processes always incline in the direction of the longest survival." — Man, The Unknown.

These statements by one of the great scientists of the age, so far as living organisms are concerned, mean much more than many understand. By reason of the perfection of the body and of the power of vital adjustment possessed by it, a power little understood by the greatest scientist, man has been able to survive through the ages in spite of his changing environment and his evil habits. The toll taken through the ages by the evils and dangers thus thrust upon the body, has been a deterioration of the body's organs and structures that has reduced its duration from a period that probably once covered nearly a hundred thousand years to that of much less than one century in Modern times.

Misleading The Multitude

Modern institutions attempt to conceal the facts of human degeneration by deception, by the destruction of ancient records, and by leading people to believe that the race is progressing, moving upward, and that man now stands at the very pinnacle of his earthly career, that his scope of knowledge is greater than the race has ever before possessed.

There is not a scrap of evidence to support these claims. They are empty and misleading. They fall before every impartial investigation, and do much damage and no good. They weaken all desire to live a better life and discredit those who present the facts. They constrain the multitude to regard and treat as an enemy any teacher who would show the way back to the better life, making it dangerous to expose fraud and advocate race improvement. The elements of Time and Tear have no effect on the living body. The turning of the earth on its axis means nothing to a body that is repaired and renewed every minute of its existence. The body neither ages nor wears — but it weakens. The causes of the weakness include man's hostile environment, his harmful habits, and medical treatment.

Carrel seemed to disregard habits and medical treatment. He saw only "the difficulties and dangers of the outside world." He is right in holding that a hostile environment is enough to weaken the body and send it to the grave, but the difficulties and dangers of man's habits and medical treatment are often more injurious than those of the external world.

As a result of the inimical conditions to which it is subjected, and the injurious substances that enter its receiving chambers (air organs and stomach), the body changes instead of wearing out, as Carrel says.

The body changes may be briefly considered under three main heads, viz., 1) a change in the organs and glands ruled by the Law of Vital Adjustment 2) causing changes in their functions, and 3) a change in the quality of the cells and tissues.

By this count, the body gently sinks into a slow process of degeneration instead of dying suddenly, giving the body maximum duration under the circumstances, because it is able to change as Carrel says. 1. The changes in the body's organs and glands, under the Law of Vital Adjustment cause some of them to fall below par and, in time, lapse into a state of dormancy or semi-dormancy. The others must increase in size and function to compensate for this loss, putting a strain on them as the body struggles to survive under the handicaps it is forced to face. 2. The change in function resulting from the change in organs and glands are numerous and are for the worse. A condition of slow deterioration sets in, and its symptoms are what doctors are trained to term "disease." In time these changes appear as diabetes, Bright's disease, arthritis, neuritis, lumbago, rheumatism, and so on, through the entire list of diseases. 3. The change in the quality of the cells and tissues appear in that condition of weakness known as "old age," where and when the activity and elasticity of youth are succeeded by the slowness and stiffness of decrepitude.

Harmful Practices From Birth

The changes are not the work of Time and Tear, but of harmful habits, a hostile environment, and medical treatment which begins after the baby is born, with vaccination and inoculation to make the body "disease-proof," and is continued all through life. The body is poisoned and weakened at the start, begins its decline into degeneration immediately, and is never given an opportunity to recover.

As the organs and glands become dormant or semi-dormant, the body and its functions change in its adaptation to evil influences that reduce much of its integrity and efficiency, and it loses much of its spiritual capacity and faculty to function on the spiritual plane.

The evidence of this loss is most apparent in the brain, and proof of this loss is found in the institutions of civilization that are filled with the insane and feeble minded. It is common knowledge that civilized man is mentally only ten percent of what he should be, and the few on earth still sufficiently sane to see the light are those who are jailed and liquidated on charges of obstructing social progress by exposing the social pattern.

Man as a Breatharian received all his substance of the Cosmos directly through his Air (Spiritual) Organs, and thus functioned on the Spiritual Plane while dwelling in the Material World. His spiritual capacity and faculty to function on the Spiritual Plane failed in direct ratio as his body changed and became more material through the internal changes suffered by his organs as a result of the body's adapting itself to the evil influences which it could not control, and which it was forced to endure or die. In the perfect state of Breatharianism, man's body was free of the clogging and dulling substances of the material world; and it must be free again, of these damaging substances before it can return to its perfect state, when man was competent to explore with his mind the Spiritual Universe and possessed actual knowledge of the fact that he is of that Kingdom and has eternal life. But one should not court danger by trying to return to Breatharianism too quickly.

In his present, degenerate state, man's body is afflicted with all the cumulative effects of his harmful habits, his hostile environment, and the poisons introduced into his body by the doctors. That is the change, as Carrel called it, which we know as the shocking stage wherein man's physical organism sinks into the decrepit stage termed old age. He is taught to expect it, as his body wears out. On the contrary, the best doctors declare that the body does not and cannot wear out, and is so perfect, as a machine, that it should go on forever.

Fewer Centenarians

As evidence of the progress of degeneration, Royal S. Copeland, M.D., former New York Health Commissioner, said in the press of June 27th 1980: "Fifty years ago there was a population of a little more than fifty million people in the United States, 4,000 of whom were centenarians. At the present time, with more than double the population of fifty years ago, there are only 2,841 people who have reached the age of 100."Dr. John Harvey Kellogg wrote: "Civilized man is dying. This melancholy fact is recognized by all students of anthropology. Such eminent economists as Major Darwin, son of the famous Charles Darwin, and Professor C. B. Davenport of Carnegie Institution, consider the case hopeless, and believe that man will ultimately become extinct through degeneration." - Good Health, August 1930.

Max Heindel, an author of note, in writing on "Man's past evolution constitution and future development," attempts to show in his book under the "Science of Nutrition," that decrepitude arid death result from a change in the body's organs, structures and tissues because of what man eats and drinks. He ignores "the difficulties and dangers of the outside world," which Carrel saw, and failed to notice the wholesale manner in which the race is being poisoned. Heindel wrote: "There is a gradual increase in density and firmness of bones, tendons, cartilages, ligaments, tissues, membranes, the coverings and even the very substance of the stomach, liver, lungs and other organs. The joints become rigid and dry. They begin to crack and grate when they are moved, because the synovial fluid, which oils and softens them, is diminished in quantity and rendered too thick and glutinous to serve that purpose. The heart, the brain, and the entire muscular system, spinal cord, nerves, eyes, etc., partake of the same consolidating process, growing more and more rigid. Millions upon millions of the minute capillary vessels which ramify and spread like the branches of a tree throughout the entire body, gradually choke up and change into solid fiber, no longer pervious to blood.

"The larger blood vessels, both arteries and veins, indurate, lose their elasticity, grow smaller, and become incapable of carrying the required amount of blood. The fluids of the body thicken and become putrid, loaded with earthy matter. The skin withers and grows wrinkled and dry. The hair falls out for lack of oil. The teeth decay and drop out for lack of gelatin. The motor nerves begin to dry up and the body's movements become awkward and slow. The senses fail; the circulation of the blood is retarded; it stagnates and congeals in the vessels. More and more the body loses its former powers. Once elastic, healthy, alert, pliable, active and sensitive, it (changes and) becomes rigid, slow, and insensible. Finally, it dies of old age."

Medical art says the body wears out. Carrel says it changes. Given above is a good description of the more apparent changes. From the changes listed many others occur that are less noticeable and harder to describe. It is unnatural for the body thus to change; and the changes do not come of their own accord. The body battles against them, but they are forced upon the body by the persistence of a hostile environment and man's harmful habits.

The Body Fights Against Changes

The changes come slowly, gradually, steadily, while the body is constantly struggling against the various handicaps in its fight to live. The evidence of these struggles are the symptoms termed "disease." The cause of the changes are not in the body, but "in the difficulties and dangers of the outside world," and in the difficulties and dangers of man's evil habits.

Rudimentary Organs

The great Carrel, in his work, "Man The Unknown," wrote: "The body seems to mold itself on events. Instead of wearing out (dying), it changes" (degenerates) (P. 192). We have described some body changes in the foregoing pages, showing how and why the body sinks into degeneration and death. There is another phase of the subject to which we shall now refer. On page 197, Carrel said that the body's intelligence possesses both a prevision and a provision. It perceives the remote and the near; the future and the present, and provides, by definite changes, for such conditions and emergencies as its prevision shows that it must meet — or perish. Carrel failed, for some reason, to proceed from there and present some evidence of changes which the body makes and has made in order to survive under new conditions that prevailed not at the time when man first came into physical being. Had he done so, he would have uncovered some strange and startling things.

Could Carrel, as a doctor, see the present rudimentary organs and glands in the body, and not understand that they must have been useful and functional at some early period in man's life? These rudimentary structures must represent changes the body has made in its struggles to survive under adverse conditions. Could Carrel view the stomach and intestines in modern man and not realize they were rudimentary structures in the days when man was a complete breatharian?

With that evil disorder Constipation so prevalent that it is termed the "national disorder," and with few folks free of hemorrhoids (piles), of stomach and bowel troubles, with thousands suffering from appendicitis and many dying from appendectomy, could Carrel pass these glaring facts and not perceive that something must be wrong? How could he, as a doctor, fail to understand that all this misery and these disorders indicate a shifting of the body from its original course? Even a layman knows that if a machine fails to perform efficiently a certain work, that failure is evidence that the machine is asked to do what it was not made to do.

Supreme Intelligence equipped the body, in its physical beginning, with all the structures that it would ever need under all reasonable circumstances. It was made perfect and complete. But it seems that man has strayed even farther from the true path of life than was ever anticipated by an omniscient Creator. Modern man has the rudimentary breast of the female. In some cases, they are functional, and such men can nurse babies, as shown by Clements in his Science of Regeneration.

He also shows that the male glands of generation appear in rudimentary form in woman, and vice versa. These rudimentary structures represent changes that have occurred in the body through the ages because of changed conditions. As a Breatharian, man had all the organs, both functional and functionless, developed and undeveloped, that he would ever need as he drifted down the stream of degeneration.

He fell from the plane of perfection by becoming a drinker of fluid and a consumer of food; thus, creating unnatural wants and desires that have dragged him down to misery and despair. He had become extinct but for the prevision of an omniscient Creator, who provided with rudimentary structures for just such emergencies. As rudimentary organs were needed and commanded into use by conditions, new environment, and new habits, they responded to command and developed to a functional degree. Thus, the body cha instead of dying. Huxley and Darwin declared that the rudime organs in men and women are the remains of structures that have better developed in an earlier state of human existence. They ar anatomical remains of what has been, and are used by modern biolo in tracing lines of descent with modification, and in determining pro ancestry.

Huxley said, "Either rudimentary or vestigial organs are use, in which case they should have disappeared; or they are of u which case they are arguments for 'telegony,' which means that the of past and future service or purpose" (Anatomy of Invertebrates, P.

Darwin made a deep study of this matter and he wrote: complex organ in a rudimentary state is direct evidence of its having been functional, and in order to discover the many transi grades through which it has passed, we must look to very ancient which have long since become extinct. Rudimentary organs of trifling importance, have probably been of high importance to an progenitor, and after being perfected in a former period, have transmitted in a more or less changed condition by modified descen until of slight or no use. In all species, or varieties, correlated vari play an important role, so that when any part has been modifi changed, other parts have necessarily been similarly affected or mo — and so viewing it, Nature may be said to have taken pains to r her scheme of modification by means of rudimentary o embryological and homologous structures, but we are too bli understand the true meaning of them" (Variation of Species, pp. 14 178).

In Lesson No. 7 appears the story of a woman who is wo back to Breatharianism. She said: *"I have passed the eating stag could not eat even if I desired, as my digestive apparatus has ch considerably, and is now unable to handle any fibre at all."*

Because it has no work to do, this woman's alimentary tr shrinking back to its original rudimentary state as it was before began to eat. In the beginning the alimentary tract was rudimenta are the mammary glands now on man's breast, and it reverts original condition when man, by not eating, gives it a chance to change back.

These are some of the changes in the body to which Carrel referred but failed to describe. For the body, it either changes or perishes; it either adapts or expires. So the body meets the emergencies by making the necessary changes. But Carrel was too materialistic to use his Mind and find within the kingdom of God (Luke 17:21) the facts he needed to aid him in discovering and describing how and in what ways the body had changed. The body, being subject to changes, can change in practically all directions to meet many emergencies and survive, even the changes that decrease its efficiency and diminish its duration.

The body can change to meet adverse conditions, and it can also change to meet favorable conditions. These changes are possible because of organic and functional changes within the body. As the organs change, their functions must change to a corresponding degree.

To describe these changes definitely and in detail would require observations covering thousands of years. Some surprising changes can occur in one generation. As where a man turns into a woman and vice versa. If a man can change into a woman so completely as to become the mother of a baby, or a woman can change into a man and become the father of a child, as Clements shows in his Science of Regeneration, then it should not seem so surprising or impossible that a breatharian can change to a gluttarian and vice versa. As a man becomes a woman, as sexual changes occur because of organic and functional changes taking place within the body, so a breatharian becomes a gluttarian for the same reason. The sex glands of a body of nine or ten years of age are in a rudimentary stage as a rule and not competent to function in a productive degree. But, as they are commanded into use by the boy's habits, they respond and he becomes competent to produce offspring.

The press of July 15th 1951, reported the case of a ten-year-old girl in Picayune, Misissippi, who gave birth to a seven pound son. The doctor said the baby was "perfectly normal." When similar reports come out of India, we in the U.S.A. think it is terrible. When it occurs in our own country, it is winked at and forgotten. Prejudice is a powerful influence.

Lesson No. 6
Man's Natural Home

It stuns a man to tell him eating is not natural. Most men never heard of people who live without eating.

We show in our work, "The NUTRITIONAL MYTH," that eating is an acquired habit, like smoking. The sensation of hunger rises from certain stimulation of the alimentary tract. "Appetite comes with eating."

An advanced scholar writes that there was not an Ascent of Man but a Descent of Man, and this theory is supported by ancient records and legends. Man did not spring from the slime of the sea or from a worm in the ground, as science claims. He came from another planet or star, travelling to this planet in a space ship, now called "flying saucers," several of which have been seen since 1947, and some have landed on earth and dead men have been found in them.

This man did not eat, but subsisted on cosmic elements. He was a Breatharian, and assimilated sunshine and cosmic rays from the atmosphere of the earth to which he had come, and acclimatized himself to its atmosphere and his new environment. In that distant age man dwelt in high places where air is purest and highly charged with ozone and cosmic rays. This cosmic substance he inhaled, and it was termed the Breath of Life. By it his body was animated and sustained. In the high altitude the weather was perpetually cool, but his powerful vitality kept him comfortable. In that day, according to legend, man had a life-span of nearly a hundred thousand years. He did not know somatic death according to Bagget Irand, who said: "During that time it was common to find men and women who were thousands of years old. In fact, they did not know [somatic] death. They passed from one accomplishment to a higher attainment of life and its reality. They accepted Life's true source, and it released to them its boundless treasures in a never-ending stream of abundance." — Life &Teachings of the Masters of the Far East, Volume II. Long ages pass and the time came when man decided to use pure rain water. So he added liquid to his sustaining substance. This man was blonde in complexion, had sparkling blue eyes that resembled the color of the sky and hair of golden yellow that resembled the sunshine.

The ancient Greeks had a tradition of the Hyperboreans who dwelt in the mountains in a land of perpetual sunshine and ate only fruit, but originally, like the gods from whom they descended, subsisted on air and sunshine. They were never ill and the duration of their life was a thousand years. The word Hyperborean means beyond or in the mountains. Man's traditional "Fall" occurred when he migrated to lower levels, where he found fruit growing and ate thereof — an event symbolized by eating the apple. Only after man descended to the low regions of the tropics, where he found fruit growing in abundance, did he become a consumer of food and darker in color.

Altitude Is Beneficial

Science shows that climate and altitude govern man. Each race harmonizes with its environment. In the high, cool regions, in the warmer middle regions, in the low hot regions, the type of people differ, but in each region they are basically similar. According to climate, altitude and the condition of the air, so is man. By these he is ruled, his constitution formed, and his habits shaped.

Regardless of where or how man lives, his body is basically composed of and sustained by cosmic rays, either directly or indirectly, in the form of minerals condensed from the rays after they strike the earth's atmosphere. It is for this reason that in high altitudes, where cosmic rays are stronger, the air contains more minerals to sustain the body, making it easier to subsist on cosmic rays at high altitudes than in low regions.

In low, warm regions people are languid, listless, of low vitality and poor health generally, regardless of how they live or the kind or amount of food they eat. The air of such regions lacks freshness and vigor; it contains too much carbon dioxide and too little oxygen and ozone. Also, the humid decomposing humus in the soil emits odors of acid decay that further weaken the body and shorten its duration. The worst air, speaking generally, is the stagnant, stifling, warm air in low regions of the temperate and tropic zones. In the latter region occurs the lowest human degeneration, and in some of these regions the average life-span is surprisingly short.

Languor, listlessness, weakness, and poor health come when the body cells are saturated with acids that disturb their mineral balance.

They lack the capacity to receive and register cosmic radiations properly. When the mineral deficiency advances far enough, the organic radio fails to function on the life level, and that state is termed "physical death."

Lesson No. 7
She Eats Nothing

In regard to eating Judith C. Churchill wrote: "When you overeat one day, you are hungrier the next. Huge meals stretch your stomach and throw your appetite out of proportion. Conversely, the less you eat the less you want. . . . After you become used to smaller food intake, you may wonder how you have previously eaten so much." - Readers Digest.

What man has done, man can always do. There is an ancient tradition to the effect that the first men did not eat, and a London lady is trying to prove it on herself. The London Sunday Chronicle of June 17th 1951, carried a picture of Mrs. Barbara Moore Pataleewa, of London, with her story that her "diet" consists of air, sunshine and an occasional glass of water. The account states: "A woman of 50 who looks like she was only 30 claimed yesterday that she hates food, has beaten old age and expects to live at least 150 years. She has set out to do it by giving up eating. "Twenty years ago she ate three normal meals a day. Slowly for 12 years she reduced her eating until she was keeping fit on one meal a day of grass, chickweed, clover, dandelion, and an occasional glass of fruit juice. Five years ago she switched entirely to juices and raw tomatoes, oranges, grasses and herbs. Now she drinks nothing but a glass of water flavored with a few drops of lemon juice to kill the taste of chlorine."

(NOTE-Killing the taste of chlorine in the water does not remove that poisonous substance from the water, and in time the cumulative effects of the poison will appear in some ailment if she continues drinking that kind of water. — Klamonti).

"She says, 'There is much more in sunlight and air than can be seen by the naked eye or by scientific instruments. The secret is to find the way to absorb that extra — that cosmic radiation — and turn it into food; that is what I have done. Every year she goes to Switzerland for the purer air and climbs the mountains on a diet of water from the streams, 'You see,' she explains, 'my body cells and blood have undergone a complete change in composition. I am impervious to heat, cold, hunger or fatigue." She continues: Winter or summer, even in Switzerland, I wear nothing but a short sleeved jumper and skirt. In cold weather people stare at me. But while they shiver in furs, I am warm. I am as strong as any man, and need only four or five hours' sleep for mental relaxation.

Because I have no toxins in my system, I am never ill. I had to advance gradually from vegetarianism to uncooked fruit and then to liquid food. Now I am struggling towards Cosmic Food. I have passed the eating stage and could not eat if I desired, as my digestive apparatus has changed considerably and is now unable to handle any fibre at all.

"Instead of thinking that my normal physical life will end in ten years, I am growing younger. With patience anyone can do the same. The tragedy is that eating is considered one of the pleasures of life. To stop eating is to experience discomfort while the body is adjusting itself to the new course. I now find the very smell of food disgusting."

Vegetarians find the smell of flesh (meat) disgusting. If they have been vegetarians long enough, to eat flesh would make them sick. We know by this and other experience that when the Breatharian first attempted to eat, it made him ill, as the first cigarette makes the youth ill. Eating still makes man ill while fasting restores health.

The statement that this woman is comfortable while thinly clad in cold weather while others shiver in furs, proves the correctness of the ancient tradition that ancient man dwelt in high regions where the weather was perpetually cool, but was kept comfortable by his powerful vitality.

She says that her body cells and blood have undergone a complete change in composition, making her impervious to heat, cold, hunger or fatigue. She further says that her digestive tract has changed considerably and is now unable to handle any fibre at all.

Survival Is Nature's Goal

The great Carrel devotes an entire chapter to the subject of Adaptive Functions in his work Man The Unknown, stating that the body seems to mold itself on events, and "instead of wearing out, it changes" (P. 192). He continues: "Our organs always improvise means of meeting every new situation; and these means are such that they tend to give us a magnesium duration. The physiological processes always incline in the direction leading to the longest survival of the individual."

We have an example of some of these changes occurring in the body of this woman. Her alimentary tract is shrinking to its original rudimentary state as it was before man began to eat. In the beginning the alimentary tract was rudimentary, as are the mammary glands now on man's breast, and it reverts to its original size when man gives it a chance by not eating. This woman is proving in her experiment that it is regular, under proper care, for the body to regenerate and return to its original perfect state. As degeneration is a fact, regeneration is a possibility.

Breatharianism To Gluttarianism

As a Breatharian, man's alimentary tract was rudimentary and his lung capacity was much larger than now. The lungs decreased in size as eating forced the development of the alimentary tract and reduced the capacity of the air organs, because eating reduced the body's need for "cosmic food." From its original state of Breatharinism, the body has gradually changed through the ages and declined to its present state of Gluttarianism. The Body has changed from a Superior Entity that was made to subsist of Cosmic Substance to an Inferior Entity that subsists largely on the gross products of materialistic substances.

The body never wears out, as is claimed. It changes says Carrel. The change to which Carrel refers is the natural way that the body sinks into degeneration from misuse and abuse, which includes bad environment and all of man's bad habits. As the body sinks in degeneration, its vitality is rendered too weak to fight to the death against dangerous and destructive conditions. Instead of going down in sudden death, the existence of the suffering body is prolonged.

It is a "change" under the Law of Vital Adjustment by which man escapes from early death for days of misery that are pitiful for the sufferer. As the body changes to adapt itself to the downward course, it must also change to adapt itself to the upward course, as shown in the case of this woman. As the body sinks in degeneration under abuse, so will it rise in regeneration under proper care.

Buried Six Months And Lives

The body is so plastic that it readily yields to man's desires and practices, whether good or bad. This is shown by the Yogis of the East who are reputed to have achieved almost unbelievable powers, by concentrated exercises and by systematic control of breathing.

Man's body is composed of soft, alterable substances, susceptible of surprising changes in function, making it last longer than if made of steel. Not only does the body last, but it ceaselessly overcomes the difficulties and dangers of its environment and of man's bad habits.

The press of July 26th carried the account of a Yogi of India, novice, who had established a record by living in a state of suspended animation for six months in a grave, without food or drink.

When he emerged from the grave, at Benares, his clothes were said to have been worn away and his body covered with white ants. By rigid body discipline he was said to have forced his beard to stop growing, and his whiskers were no longer than when he was interred.

The statements were made by Dr. B. L, Atreya, professor of philosophy in the Benares Hindu University and general secretary of the Indian Society of Physical Research. The yogi, practicing the art which consists in suppression of all mental activities, discipline of the body, control of involuntary muscles, and a few other weird things, lay in the grave from September 25th 1941, to March 21st 1942, the doctor said.

Spent Time In Cell

The yogi spent his time in a pit cell, reinforced with brick and cement. The day he emerged from the tomb a crowd of more than 100,000 persons was on hand, the doctor stated.

An opening was made in the outer enclosure and then an opening in the all-around closed cell, the first slab of stone was removed by selected persons, some of them Hindu professors and college teachers.

Said Dr. Atreya: "The yogi was already awake, and he raised his hand to indicate that fact. Then he was dressed in new clothes, His old clothes were partly worn away under the influence of the atmosphere inside the pit and partly eaten up by white ants, some of which were found collected over portions of his body. He was then brought out covered with blankets and placed on an easy chair on high platform, visible to all. The yogi looked just the same as when he entered the pit, Even the beard on his face had not grown. He insisted on walking about 25 yards from the place, but we did not allow him to do it for fear of his being crushed by the crowd which wanted to get close to him and touch his feet." — Grit, July 26th 1942. It appears that the Yogi know certain secrets of the body and its function which enable them to suppress all mental activities, control the involuntary muscles, and withdraw the senses from natural outward expression.

Lesson No. 8
Materialism

The Materialist missed the point when he wrote: "God formed man of the dust of the ground.... For dust thou art, and unto dust shall thou return." — (Gen 2: 7; 3: 19). As a block of ice is invisible cosmic substance materialized, so is the body of man.

Fasting experiments prove that vitality, heat, minerals, etc., come not from what man eats. They are cosmic rays that materialize as visible substance. As cosmic rays condense and become visible as matter, they lose none of their properties, one of which is vital force.

Science says that vitality and heat come from the combustion of carbon compounds in food, according to its calorie theory, which erroneously compares the living organism to a steam engine

Lakhovsky held that the living cell is an electro-magnetic entity, activated by cosmic rays, the source of its vitality. The cell's development is also directed by cosmic rays, which materialize in the form of body minerals and create the vital condition needed for the synthesis of atmospheric nitrogen into body protein, as well as atmospheric carbon dioxide into body fat. It appears that four-fifths of the air we inhale is nitrogen. The body synthesizes this substance to form protein, which occurs by its union with hydrogen in the alimentary tract. The body cannot use the nitrogen of protein foods. Practically all protein nitrogen that man eat is eliminated in the form of metabolic end-products.

Weight And Vitality Loss Due To Autointoxication

During a fast the body loses weight because it is toxic. There is a condition of autointoxication, and the internal toxins liberated cause decomposition of body protein and fat. If the organism were sufficiently pure, no condition of autointoxication and no loss of weight would occur during a fast. Inhale the terrible odor from the body of a person who has fasted eight of ten days and you will think his body is rotten.

When one's vitality decreases as one stops eating, it is due to auto-intoxication which then begins, and not to lack of food to supply energy. Food supplies no energy. Lakhovsky showed in his experiments that protein and other substances occurring in the body are converted from cosmic rays by the body's physiological processes. Organic growth and maintenance, says Lakhovsky, are the, work of cosmic rays. The living organism is a materialization of these rays. They are subtle streams of substance of ultra-electronic and materialize into grosser minerals as they strike the earth's atmosphere. So the body is a materialization of "cosmic food." He demonstrated this fact by keeping unicellular organisms in sealed test tubes, measuring the amount of iron they contained before and after a certain period of growth. He found the amount of iron increased as the cells multiplied, even though the test tubes were sealed.

The extra iron came from the cosmic rays to which the cells were attuned. They absorbed these cosmic rays at an iron rate of vibration and the rays materialized as iron atoms, showing that the cells of the body are maintained by cosmic rays. Babbit showed that sunlight is converted into minerals in the body, according to the spectral colors, each corresponding to different groups of minerals. All that food does is to furnish a certain type of stimulation. As we advance from vegetarianism to liquidarianism, and then on up to Breatharianism, we change from the grosser forms of stimulation to the finer forms. Instead of the body getting its stimulation from food and liquid, it gets its stimulation from the elements in the air.

When the scientific theory of Materialism exploded, material science exploded with it and its textbooks became obsolete. Professor J. S. Haldane, the great astronomer, said: "Materialism, once a plausible theory, is now the fatalistic creed of thousands (of physical scientists), but materialism is nothing better than a superstition, on the same level as a belief in witches and devils. The materialist theory is bankrupt."

With the discovery that atoms are composed of electrons and protons, and that these elements are merely whirling centers of force in the ether, material science saw its fundamental theories swept into oblivion. Physicists and chemists now know that all Matter is vibratory electro-magnetic activity. Matter is composed of units called atoms and atoms are composed of varying numbers and arrangements of electrons and protons, which are tiny centers of vibratory activity in the ether. Each of these vibratory centers possesses a magnetic polarity that is positive in the proton negative in the electron. All matter is fundamentally the same basic substance. The different properties which distinguish the various types of matter in the human body, such as proteins, carbohydrates, fats, etc., are, basically, nothing more than the differences in the number and arrangement of the protons and electrons.

The protons and electrons are held in their orbits in the atom and regulated as to the combinations they form, by the field of electro-magnetism generated by their rapid motions. In order to transform any given type of matter into another, as proteins into fats, it is necessary only that the vibratory frequency of the electro-magnetic activity composing the matter be altered appropriately. The transformation of matter is accomplished by exposing it to vibratory activity of the appropriate frequency, impelled by a force greater than the force impelling the vibrations of the substance to be transformed.

A simple illustration of this process appears in the transformation of ice into water and water into steam by exposing the substance to heat. The normal vibratory frequency of the substance composing the ice is increased by heating until the substance assumes a gaseous form, and the ice is transformed into invisible elements that float in the air.

Live Without Eating

The press of January 31st, 1981, said: "Authentic reports from Salisbury, South Rhodesia, state that Mrs. A. C. Walter, a noted singer, has been fasting 101 days, during which time she has consumed only two to three pints of cold and hot water daily. Last October she weighed 232 pounds, so she decided to fast. She has lost 63 pounds and says that she is in perfect health, goes out to parties, and carries on with her public singing."

Bernarr Macfadden of Physical Culture fame reported a case where he fasted a man for 90 days. He wrote: "The man lost 75 pounds during this period. He weighed 300 pounds when he began this fast, and 225 pounds when the fast ended."

Macfadden adds: "If a bear can fast all winter, there is no reason why a man could not do the same thing."

The Great Body Normalizer

No measures known will so surely, safely and speedily normalize a deranged body as will fasting. It is the most natural and certain of all remedial procedures, it stops at once the introduction into the body of all new material, except air and water, thus releasing the organs from the labor imposed by eating, and giving them an opportunity to purge the body of the internal poisons responsible for illness.

On April 28th, 1929, Paul Urban, a German world war veteran and professional nurse, ended a 64 day fast, during which time he took a pint and a half of pure water daily. He weighed 165 pounds and dropped to 113 pounds. He stated that fasting rejuvenates the body and makes man live longer. He was 46. In the press of July 19th, 1929, appeared an announcement of cancer being "cured" by fasting, with a picture of the patient and his nurse, under which was this statement: "Albert Schaal, age 58, shown as the flax king of Manitoba, Canada, after a fast of 49 days under the direction of Dr. Harry C. Bond of San Francisco, is said by doctor to be cured of cancer."

In his "Believe it or Not," in the press of January 25th, 1938, Robert Ripley stated that for ten years Giovanni Succi travelled through Europe living exhibitions of fasting. His exhibitions, rigidly controlled, extended for periods of 30 to 40 days. During that time, he was in the public eye day and night. Included were 80 periods of 80 days of fasting, and 20 periods of 40 days of fasting — a total of 3200 days without eating, or eight years and 280 days without food in ten years. In his "Believe it or Not," in the press of January 16th, 1934, Ripley stated that Jekisiel Laib, of Grodno, Poland, fasted six days week for 80 years. Each Saturday he ate bread and water. His health was good. According to the dietetic experts he should have died of "mineral starvation."

According to the press of July 26th, 1942, a Yogi at Benares, India, was buried in a grave for six months without food or water. (Report given in this lesson). The press of November 30th, 1934, reported the case of a Jain priest, Muni Shri Mierilalji, of Bombay, who fasted for 259 days, taking nothing but water. He ended his fast in the presence of 500 co-religionists. The press of October 12th, 1948, reported the case of a British girl of 12 years who fasted for 18 months, taking nothing but water. The press of February 6th, 1937, quoted Mrs. Martha Nasch, age 44 of St. Paul, Minnesota, as asserting that for seven years she had eaten nothing, and affirmed her willingness to submit to surveillance to prove her claim. The press of May 31st, 1948, reported the case of a Chinese girl who had eaten nothing for nine years.

The case was reported to Dr. T. Y, Gan, of Chungking Municipal Hospital, and he went to see her. Her name was Yang Mel, she was 20 years old, weighed about 85 pounds, and led a perfectly normal life, except for not eating, and drinking very little water, she showed no signs of starvation, and appeared no different from other girls, Gan said, "1 found it difficult to believe her story." The girl was never hungry, had no desire for food, and never asked for any. When asked as to why she did not drink more, she said that it made her feel uncomfortable. Her alimentary tract was so dormant and rudimentary that it could not take water without bad reaction.

In an article entitled "Forty Years Without Food," N. P. Chose wrote: "Caribala Dassi, sister of Babu Lamboxar Dey, a practicing pleader of Purulia, has been living for the last forty years without taking any food, not even water, and has been doing her regular household duties with no apparent injury to her health. Many respectable persons can testify to the truth of this statement. "-India's Message, January 1932. According to the press of May 27th 1987, Sirmati Bala, of Bankura, India, age 68, had touched neither food nor water since she was 12 years old.

One case is sufficient to show what is possible in a million other cases. Biologists are being convinced that eating is an acquired habit, like smoking, and a pleasurable indulgence rather than a physiological necessity. It is said that in India certain sects of yogis live without eating, and in the Himalayas there are many who consume no physical food.

Lesson No. 9
Body Building Material

All textbooks on anatomy teach that the human body is a composition of trillions of cells. The cells are not composed of food.

Science admits that the Parent Cell is not the product of food. It also admits that all the subsequent cells are not the product of food. It is law that what food does not and cannot produce, it does not and cannot preserve and sustain. To speak or write of cell nutrition and body nourishment is not only unscientific, but an admission per se of anatomical and physiological ignorance, even though the statements may be those of a great doctor. When health officers die as they do in their fifties and sixties, it shows that they do not know what they should.

One author says that we have one foot mired in an antiquated medical system that is dying, and with the other foot we are holding down a modern health system that is struggling to be born, which advocates Health by Healthful Living in harmony with God's Law of Life. The origin and work of the Parent Cell is a mystery; that is the cell which begins the building of the body. That cell comes not from the parents.

The so-called seed of the parents is not seed in the sense that it produces man. What is considered as the seed appears to do nothing more than to form a central, electro-magnetic point, around which occurs a condensation of invisible substance from the Cosmic Reservoir. The food one eats neither form nor sustain the body cells. Nor do they come from the, Parent Cell, which is just an electromagnetic center of crystallization and materialization, a pattern, around which Cosmic Rays materialize into cells that form the growing body.

In referring to this mystery, the great Carrel wrote: "The body builds itself by techniques very foreign to the human mind. It is not made of extraneous (foreign) material, like a house. It is composed of cells, as a house is of bricks. But it is born from a cell, as if the house originated from one brick — a magic brick that would begin making other bricks (of material that seemed to come from nowhere). Those bricks without waiting for the architect's drawings or the coming of the bricklayers would assemble themselves and form a complete house... as does the body and all its various parts" (Man The Unknown).

Whence come the cells? or the materials of which they are composed? They rise as shadows and become substance as the result of the condensation and materialization of Cosmic Rays. That substance is not food. It constitutes the elements of the Universe that have always existed and are eternal. As the so-called seed of the parents come near to each other, certain elements of each stand out separately and, coming nearer, these separate, individual particles merge and fuse as it were into each other, producing a clear field in which nothing appears. Finally, after a period of seeming quiescence, granulation occurs at a point between the places occupied by the gametes of the parents when merged, fused and disappeared from sight (P. 194).

When the so-called seed of the parents meet and fuse, they thus create a condition or electro-magnetic center, which is necessary for the occurrence of the phenomenon that produces man's body. That center attracts cosmic rays of a definite frequency, corresponding to the chemistry of that center. The rays crystallize around the center in the form of similar substance, and man comes into physical being under a magic process of transformation of invisible elements into visible form.

The great Carrel missed the point when he said that the body is not built of "extraneous material, like a house." The infant body must receive for its growth material from some extraneous source. That source is the invisible cosmic rays. The process of growth is the work of these rays as they materialize into blood, bone and flesh.

Discovery Amazed Material Science

A block of ice represents a materialization of cosmic vapor. So the body of man represents a materialization of cosmic rays. The one process is as simple, as complex and as mysterious as the other. Both processes are ruled by the same cosmic law. The scientific theory of cell reproduction cannot explain how the Parent Cell produces cells that form blood, bones, muscles, nerves, heart, brain and other organs. These differentiated cells each represent cosmic rays of different wave length.

The cosmic rays become visible by condensation and materialization of Invisible Elements that exist only as vibratory waves, whirling centers of force, termed electrons, concerning which Dr. H. H Sheldon, University of New York, wrote: "Electrons, long regarded as the ultimate particles of which all Matter is formed, have now been shown to have a reality only as a wave form, while an atom consists of a bundle of such waves." This discovery amazed material science and exploded the basic theories of Materialism. For it shows that the cells of the body consist of bundles of waves, not of assimilated food. So man is not what he eats. The body cells are composed of atoms only, and the body is constituted of nothing but bundles of vibratory waves.

So far as man's physical structure may be concerned, this doctrine might be contested by the old school scientists, but Sheldon says: "We as individuals undoubtedly have no existence in reality other than as waves, multitudinous and complicated centers, perhaps, in what we call the ether. We are analogous, in a sense, to the sounds that issue from a piano when a chord is struck, or when a symphony orchestra sounds."

This raises the question, WHY DOES MAN EAT? In our work 'THE NUTRITIONAL MYTH" we wrote: "Cell and body nutrition is a myth. What man consumes as food does not supply cell nutrition by assimilation as science teaches. The ingested substance merely produces activity in cell function by STIMULATION and not by nutrition. Two types of stimulation seem essential for the function of living cells: Vital and Chemical. The Vital is from the Source of Life, while air, liquid and food supply the chemical. The ingested substances contact and stimulate the cells into certain activity, and pass from the body through the eliminative channels, as flowing water turns the wheel of a mill, activating the mill machinery that does the grinding "-P. 17.

The process of organic growth from the electro-magnetic center formed by the Parent Cell, and the so-called process of nutrition, are one and the same phenomenon. Cosmic rays strike the chromosomes of our cells, which act as minute receptors of cosmic radiation, and the rays materialize in our cells into various chemical elements requisite for organic growth and maintenance. The magnetic chromosomes of the cells attract the electronic rays of corresponding vibratory frequency and they materialize in the cells as minerals.

What we term minerals are the foundation of the living cell. These minerals are substances that are universal in existence and eternal in duration. These substances are electrically charged particles of various minerals. The body is a complex of minerals, consisting of electrons combined into atoms and molecules. It is not composed fundamentally of proteins, carbohydrates and fats. The substances in the body are actually atoms of nitrogen, oxygen, hydrogen and carbon dioxide in various combinations, as water is a combination of hydrogen and oxygen. The nature of the substance depends upon its atomic combination.

Look To This Day

For it is Life, the very Life of Life.
In its brief course lie all the Verities
and
Realities of your existence;
The Bliss of Growth;
The Glory of Action;
The Splendour of Beauty;
For Yesterday is but a Dream,
And To-morrow is only a Vision;
But To-day well lived makes every
Yesterday a Dream of Happiness, and
Every To-morrow a Vision of Hope.
Look well, therefore, to This Day!
    -From the Sanscrit.

"A thousand years hence the contents of this work will be as up-to-date as at this hour . . . writings and methods of living based on Cosmic Law are always in order and never become obsolete."

Lesson No. 10
The Aging Process

"By old age of the body, that does not age; by the death of the body, that is not killed. It is thy Self free from sin, free from old age, from death and grief, from hunger and thirst." — Chandogya Upanishad.

People want to look younger and live longer. To that end many books have been written and the authors died young as proof that they were incompetent teachers. Why does man grow old? If the earth's turning on its axis does not produce "old age," what does?

An unusual account of Old Age appeared in the press of July 19th 1952. The item, date-lined Chicago, said: "A four-year-old girl, who weighs only 7 pounds is dying of old age at University of Illinois Research and Educational Hospital. The child is a victim of progeria or premature senility. Doctors at the Hospital said it is one of the rarest ailments. Both its case and cure are unknown. The child, named Linda, entered the hospital January 28th 1948, when she was two months old, has been there ever since. A hospital spokesman said Linda is withered and wizened with thin, balding hair, is only two feet long and wears doll dresses and shoes."

Nothing happens by chance. We may not always understand the working of the law because our view of life is so limited. If our thoughts penetrated beneath the surface, we would find a cause for every effect.

We are not told why the child was taken to hospital when only two months old, but the aging condition of her body is evidence to prove the harmful effects an artificial mode of living. Everything natural is banished from hospitals and the laws that rule natural phenomena receive no attention in such places. In the matter eating, Abbe N. De Montfaucon De Villars states that the Ancient Masters ate food only for pleasure, and never of necessity (Compte De Gabalis, P. 63).

We are told that in the Breatharian Age man's in its perfection, required not that kind of stimulation which physical food now furnishes.

Our fundamental concept of man should be that, as we know him, he is a degenerate representative of the original. His environment, greatly changed and adversely affected by the conditions called civilization, and his habits and practices, most of them bad, have forced the body to alter its functions in order to survive. Otherwise it had perished.

In the course of long ages the body's functions have changed, by continuous adjustment, and developed a dependency upon certain kinds of stimulation, rising from man's environment and his eating and drinking habits, that were foreign to the body in its original state, when it received directly from the Air, the cosmic reservoir of all things, the stimulation needed to activate its cells. Poverty and wants are conditions created by man's living an artificial life. The less we need the more complete we are, and we attain perfection only when free of all wants. The more wants we have, the less complete we are, and the farther we incline from perfection.

Conditions Of Artificial Life

Daily experience proves that the body still continues to adjust itself to man's additional errors, such as smoking, drinking, and eating certain things. Some find it impossible to smoke, while others cannot tolerate certain foods which some seem to enjoy. So the body has been forced, by long ages of eating, either to adjust itself to the foreign substances man eats, or perish. Instead of dying quickly as a result of man's errors, the body changes and sinks into degeneration. Our organs always improvise means of meeting every new situation, and these means are such that they tend to give us a maximum duration under the circumstances. The functional processes always incline in the direction leading to the longest survival of man (Carrel, P. 192).

So-called food is foreign to the body. None of it enters into the body's constitution and construction. If man's body were built of what he eats, a process of physical transformation would in time change the body literally to resemble physically the things man eats. If man is what he eats, if the body were built of the food man consumes, the eating of pork would in time transform him physically into a pig. The body was forced to adjust itself to what man eats in order to survive. It was either adjustment or death. The adjustment has become so complete that man now seems to "starve to death" when deprived of that stimulation which food furnishes.

Return Must Be Slow And Gradual

The return or transformation to Breatharianism, where food is no longer essential for body stimulation, must be slow and gradual. Man must slowly reduce the amount of food ingested daily in order to give the body time to meet the new condition and adjust itself to the perfect physical state of long ago, when the air man inhaled supplied all the stimulation the body needed. We must also leave the polluted air of civilization, or perish.

Some scholars assert that man has been on earth six to eight million years, and produce certain evidence to prove it. Dr. W. C. Pei, research fellow of the Chinese National Geological survey, unearthed "The Peking Man's remains near Peipine in 1929," and this discovery, according to science, pushes back man's appearance on earth fifty million years.

Antiquity Of Man

In his book titled "Sree Krishna," Premanand Bharati states that this is the 28th Divine Cycle of which the first three sections, viz., Golden Age, the Silver Age, and the Copper Age, have passed away. We are now in the early part of the fourth section, the Kali or Iron (Dark) Age.

The Divine Cycle is composed of 12,000 Divine Years, each of which is equal to 860 human years. So that 12,000 Divine Years multiplied by 860 give us 4,320,000 human years, which is the length of a Divine Cycle, as this is the 28th Divine Cycle, that would be a total of 120,960,000 years. The Hindu scriptures state that man had been on earth 4,000,000 years when the Great Deluge occurred.

The author of "Sree Krishna" says: "These men (of the Golden Age) required little material nutrition; they ate very little food, consisting of fruit only, and drank water — and these between long intervals."

Some physiologists hold that it requires three-fourths of man's time on earth to descend from Breatharianism to Gluttarianism, which was accomplished by an alteration of the body's functions and needs as it adapted itself to the new conditions and practices with which it came in contact. This would place the Breatharian state so far back in the night of time that little evidence of it could be found, other than what we learn now by fasting a man, who begins at once to regain health when given no food, and even shows signs of growing younger.

Professor Morgulis wrote: "The acuity of the senses is increased by fasting, and at the end of his 31 days abstinence from food, Professor Levanzin could see twice as far as he could when his fast began."

Bernarr Macfadden of Physical Culture fame wrote, "I have consistently maintained that the body can be revived and rejuvenated in every way, mentally, physically, etc., by fasting."

Dr. Moeller said, "Fasting is the only natural evolutionary method whereby, through a systematic cleansing, the body can restore its equilibrium by degrees to physiological normality."

Mayer, eminent German physician, declared: "Fasting is the most efficient means known for correcting disease" (The Wonder Cure).

Dr. Densmore wrote: "We find one great cause that accounts for the majority cases of longevity — moderation in the amount of food eaten" (P. 295). Dr. Evens said: "Among instances of longevity, we have the ancient Britons, who, according to Plutarch, 'only begin to grow old at 210'. Their food consists almost exclusively of acorns, berries and water."

Drs. Carlson and Kunde, University of Chicago, found that a fast of 15 days restored the tissues of a man of 40 to the physiological condition of those of a youth of 17. This amazing discovery seems to explain the biblical statements, "His flesh shall be fresh as a child's; he shall return to the days of his youth (Job 33:25). And thy youth shall be renewed like the Eagle's" (Ps. 103:5). Here is more evidence of damage to the body by food. Eat little of a simple diet and we live longer and look younger.

Stop eating and the sick man automatically and spontaneously begins to recover his equilibrium. As if by magic the disorders disappear and health returns. For that reason fasting is often feared and bitterly condemned by some. A world of health would put some people out of business. The press of February 26th 1948, stated that illness brings physicians of the U.S.A. $1,500,000 daily."

How To Reverse Physical Appearance Of Aging

The turning of the earth on its axis has no affect on the human body. Looking older as the years pass is the effect of physical adjustment to adverse conditions, and shows more complete adaptation of the body to bad habits and bad environment. Drugs, medicines, vaccines, serums and tonics are not the answer. Supply better living conditions and the physiological process of degeneration changes to regeneration, and the physical appearance of aging will reverse.

All things that damage the body age the body. If food damages the body it ages the body. Sickness begins to age the body in childhood because sickness results from damage to the body. Drugs, medicines, serums, hot and cold baths used in sickness, age the body because they damage it. Polluted air, bad water, hard water, chlorinated water, tobacco, liquor, heavy manual labor, excessive exposure to the hot summer all kinds of riotous living — these age the body, and the body improves when such practice or habit ends. Remove the CAUSE and you have found the CURE. A library of medical books is unnecessary to teach that simple law of Cause and Effect. Under no circumstances can man stop breathing and live. Every living thing must have air or die. To stop the breathing is to stop the living. Water comes next. Man may go without water for days and live. If the air is very moist, he can live longer without water than in a drier atmosphere. Men at sea, when shipwrecked and have no fresh water to drink, are able to supply the body's needs by submerging the body in the water. The salt is filtered out as the water is absorbed through the skin.

No one can breathe too much good air or drink too much good water. But one can easily eat too much of the best food and the result is always bad. No eating habit is harder on the body or causes it to age faster than that of salting food. Anyone who claims the body needs common table salt should try drinking sea water when thirsty.

Constipation The National Disorder

The practice of eating being considered as bad, is further shown by the fact that few people are free of stomach and bowel troubles, while constipation is so universal that it is termed the national disorder. This should not be if eating were natural for man. If food were not foreign to the body, it should not derange the so-called food organs, the digestive tract and its accessories. If food were necessary to sustain the body, then the fasting of patients would be dangerous and could not be the "cure all" that it seems to be. We know the body adjusts itself to many abuses and becomes accustomed to many new conditions. So it can adjust itself to feeding. But little does man know that *all such adjustments can occur only at the expense of a depression of the vital functions, which must be injurious if long-continued or often repeated. That is what the body has suffered because of eating.*

While it appears from the evidence that eating is not natural, the body as a machine is so perfect that it can take such abuse and survive for a century or more, provided the amount of food consumed is not too great.

That fact was shown in the case of Ludovico Cornaro, who was a physical wreck at the age of 40 and told by his physicians that he could not live. He fooled them by turning to Nature and recovering health to such extent that he lived to be 103. Cornaro found that a simple diet of 12 ounces of solid food and 16 ounces of fresh fruit juices daily was comparatively better for him. On his 78th birthday his friends urged him to increase his ration. Reluctantly he agreed to an increase of only two ounces of the same food. In twelve days, he was ill with fever and pains in his right side. He returned at once to the 12 oz. ration, but suffered for 35 days. That was his only illness in 63 years on his frugal fare.

One case is sufficient to show what is possible in millions of other cases. Cornaro proved on himself the virtues of frugal feeding, contrary to medical advice that man must be "well nourished." The consumption of much food to build up the body's "resistance to disease," works the other way. Looking older, old age, a state of decrepitude that appears with the years, is the result of the body's adjustment to harmful habits and adverse environmental conditions. No hunter ever found a wild animal that showed signs of old age. Supply better living conditions, discard bad habits, move to an environment of good air, keep the home well aired, practice chastity, and the appearance of "aging" will retard and reverse in time to that where the body was when its equilibrium began to show imbalance or deterioration. That takes time, as the "aging" process must first come to a full stop before improvement can begin.

Lesson No. 11
Does Man Starve

"As long as we confine ourselves to the world of observation, we must continue in a state of bewilderment." — Robert Walter, M.D.

We live in a world of illusion. We are victims and prisoners of our five senses, and they are unreliable. We are not surrounded by what we think we are, nor do we actually see what we think we see.

The press of July 23rd 1951, quotes Dr. Theron Alexander, psychologist, Florida State University, as declaring: "The familiar saying of 'seeing is believing' is being seriously questioned these days. It is more the reverse; we tend to see what we want to see."

There are two systems of Thought — the backward and inward, and the forward and outward, the inductive and deductive, the empirical and logical. Both systems are claimed to be based upon Fact, but not upon the same class of Fact. One bases all practice upon the facts of observation, and claims that its processes are inductive. This is the backward and inward process, and Francis Bacon was its great representative. The other system, the forward and outward, is based upon a fundamental truth or principle, and was followed by the Ancient Masters, who taught that the secrets of the Universe are discovered by studying Causes instead of Effects (Rom, 1:20). All observable facts are the effects of something preceding; the something which preceded is the Cause, which, being discovered, constitutes an eternal verity which changes not, and so becomes the unchanging basis from which logical reasoning may be conducted.

Professor Robley Dunglison, one of the ablest authors and professors, warns us against reliance on observation, and quotes the man who —

*"Saw with his own eyes the Moon was round,
Was equally sure the Earth was square,
For he'd travelled twenty miles and found
No sign that it was circular anywhere."*

Dunglison recited many facts to show the fallacies of medical observation and the absurdities of medical practice that have grown out of observation — out of what men think they see. Chapman, Bichat, Magendie, Bennett, Holmes, all laughed at or exposed medical reasoning based on observation.

Finding The Truth

We cannot reach truth by reasoning from facts of observation. For such facts clearly show that the sun does rise and set.

"True science is in the mind," wrote Professor Jevons. The postulate from which logical reasoning proceeds must also be in the Mind, inasmuch as Effects are only the visible symptoms of an invisible cause. This makes the Fact to be not one of Observation but one of Conception.

Empirical scientists reason not so much from what they observe as from what they THINK they observe. Appearances are usually deceptive and seem to be what they are not. So it is not surprising that Empirical Science, even though dignified by the new title Inductive, has always proved unreliable and erroneous. The world of observation is the Shadow World, not the Real World. It is the World of Effects, not the World of Causes. What we see are the visible Effects of the operation of Invisible Causes. If the body cells are self-existent and eternal, if they are not produced nor sustained by food as Carrel showed, that raises the question as to why man appears to starve to death when deprived of food.

We could well counter with the questions: Why does man die when deprived of Opium? Why does man sometimes drop dead upon the receipt of very bad news?

Food Stimulates

In our "THE NUTRITIONAL MYTH" we stated that the substance man eats does not nourish his body. It merely contacts and stimulates the body machinery into a certain state of vital activity, and passes out of the body through the eliminative channels, as flowing water passes on as it turns the wheel of a mill on the bank of the stream.

The water activates the machinery of the mill that does the work by acting on the water-wheel. But the water never becomes a part of the mill. So the food one eats activates the machinery of the body, but never becomes a part of the body. It may appear to physical science that man starves to death because he is deprived of food. This is another instance where we must not be deceived by what we think we observe. It is not yet known how electrons become combined into atoms, or how molecules become combined into cells and protoplasm to build and sustain the living body. It has become evident that the body is not built and sustained by food. Food is the occasion and condition that activates the vital processes and stimulates their activity. Unorthodox physiologists assert that such activation and stimulation as food now furnishes, were not necessary before the body altered its functions and adapted itself to the practice of eating, as it still does to the practice of smoking.

Some of the Flying Saucers we have read about since 1947, have landed on the earth from some distant planet. They are disk-shaped ships, made of metal lighter than aluminum but stronger than steel, and they move by electro-magnetic power. Frank Sully reports in his book that a group of scientists examined a Flying Saucer that landed near Aztec, New Mexico. In it they found sixteen dead men, ranging in height from 26 to 42 inches. Their bodies were normal from every viewpoint. They had a type of radio unknown on this earth, and subsisted on tablets of an unknown kind of concentrated food. In all probability that food would not sustain us because our body is not adapted to it. Thus, we might eat and starve to death on the food eaten by these men.

Carrel says that our ignorance of the body and its functions is profound. If we know so little now about the body and its functions, we have no knowledge at all as to the body and its functions in the past when man first came into physical being, in an Environment so perfect for the body's needs, that it actually drew man out of potential existence in the Spiritual Realm to actual existence in the Physical Realm.

We must assume that in that day the Environment was in perfect harmony and attunement with the body, and vice versa, or else the body had not appeared on earth as a physical entity.

It Is The Body That Acts

If man's body, in less than a generation, will change and adjust itself to a point of craving to the death a poisonous stimulant such as opium is known to be, it should be easy to realize that, through long ages of eating, the body has changed and adjusted itself to the point of craving to the death that stimulant termed "food." As an example, it is absurd to hold that the action following the administration of a drug, is the action of the drug. There is a mechanical contact of course, and there may be some chemical affinity between the elements of the drug and the organism. But the action that follows is the action of the vital system upon the drug in its process of making proper disposition of the drug.

The same principles apply to food and drink, and to everything that enters the body. It is the living organism that acts and not the ingested substance, whether it be food or drugs. There is no greater error than to assert that the body derives energy, vitality and strength from food.

People Who Crave Poison

A staff nurse in a certain hospital accepting drug addicts tells of her experience caring for such cases. She said: "If I had ever been tempted to take drugs, I had discarded the idea quickly after working in a hospital ward for drug addicts for the past two years. There can be no torment of the damned that exceeds the agonizing pains that a drug addict goes through to conquer the habit. "All our patients entered the hospital ward voluntarily, but as they were not released until completely off of the drug, it was a pitiful sight to watch them. The first day they were bolstered up by friends' and relatives' good wishes, and had up every bit of will-power that the drugs had left them. By the second or third day all will-power had vanished as the dosage of the drug was slowly reduced, and they became abject, tortured, screaming idiots as their nerves begged for the drug."

That is an example of the great power of the law of adaptation. The use of poisonous drugs that destroy the body by degrees and never do the body any good, makes man an "abject, tortured, screaming idiot as the nerves beg for the drug."

Power Of Adaptability

The power of adaptability is one of the ever-present facts of living existence. Men live in every climate, are subject to all kinds of influences, and indulge in every sort of habit. It has become proverbial that "habit is second nature", and "what is one man's meat is another man's poison."

The evidence is hourly apparent that man may become accustomed to almost anything short of shooting or hanging. No matter how repugnant or destructive a substance may be, we first endure it and then embrace it, provided time is given to secure the efficient operation of the body's Balance-Wheel, whereby is prevented a violent swaying of vital activities from one extreme to the other. It is the sudden and violent changes that become immediately destructive to the body, even sometimes when it is a change from bad to good. Because a habit does not seem to be immediately destructive it is not proof that it is either beneficial or harmless. The secondary effects are the real and lasting effects.

The bearing of the Law of Vital Adjustment on many important questions is evident. Any discussion of the subjects of food and drink, the treatment of those addicted to the use of liquor, tobacco, opium, food, involves this law. It is a lack of knowledge in this respect that has prevented the success of many reforms which had in them the elements of great value had they been carried to a successful issue. All changes in man's habits should be made with a distinct recognition of the fact, that Vital Accommodation will succeed provided time and opportunity are accorded. The Eater must not expect to become a Breather in a day nor a year. He must return to the perfect state as he departed from it.

The body must have time to change, alter its functions and adapt itself to that perfect state in which the Breatharian was when the Breath of Life supplied all his physiological needs and he was free of all poverty and want, of all discomfort and sickness. Suddenly to shift the Hottenton to the frigid region of Greenland, or the Greenlander to the hot region of tropical Africa; to force upon one totally unused to polluted air, tobacco, liquor, drugs, opium or arsenic, which seem normal to those who indulge in them; or even quickly to deprive the victims of their indulgences, causes suffering, disaster, and even death.

Move man from the low lands to mountainous regions and he experiences much difficulty in breathing, as his lungs are not adapted to the higher altitude. Keep him there and the adjustment will occur in the expansion of his lungs and the increase of their capacity.

The Breatharian lived in high regions and his lung capacity was much larger than that of the Gluttarian. Every practice makes a place for itself in the vitality economy; and suddenly to cease any practice leaves a vacuum that is uncomfortable at least.

Every fact of life, as well as the Law of Vital Adjustment, goes to prove that man is a product of Evolution — but not in the sense that one species is evolved from another. Neither in the sense that animals are grown out of vegetables, nor man from animals, except on a plan which is occurring under our daily observation. Cosmic Laws never change.

The passage of vegetal and animal material through the human body in the process of ingestion, digestion and elimination is a fact of observation. But these substances never become a part of the body, do not sustain it, and do not change its constitution.

Fish Does Not Give Brains

Eating bacon does not transform man into a hog; nor will eating grass-seed make of him a bird; nor eating fish give him brains — medical art to the contrary. If food built, nourished and sustained the body, it would its constitution. That is a cosmic law that has no exception. If a house be built of bricks, it is constituted of bricks and not of wood or stone.

The press of May 3rd, 1936, reported that a Hindu Bengali Kayastha woman of 68 had eaten nothing since she was 12. The account stated: "Swami Yogananda Giri, the disciple of Guru Shyama Charan Lahiri of Benares, came here recently from America and, accompanied by Messrs. Sandanada Sanyl, press representative, and Bibhuti Bhusan Ghosh, went to Patrasayar in the afternoon to visit the woman and gather from her the following information: She is active, discusses higher philosophy and religious matters, and an expert in 'Pranayama and Yogi.' She takes nothing, not even a drop of water. She is always gay and looks like a child despite her age. She does not pass stool nor urine and does her household work like any other woman. The Swamiji wanted to take her to America with him, but she was not agreeable to this, as she had no order from her Guru to leave her native village." — Amrita Bazaar Patrika.

One case is sufficient to show what is possible in a million other cases. Another person lives on nothing but orange juice for six months; another fasts six days a week for thirty years, eating only bread and water each Saturday. It is not disputed that the Law of Vital Adjustment modifies physical development and function. It is believed the practice of parents become constitutional tendencies in their offspring. How far this process of development may extend, no one is able to say.

Lesson No. 12
Danger Of Abrupt Changes

"The science of living beings in general, and especially the human individual, has not made much progress. . . . We have grasped only certain aspects of ourselves. We do not apprehend man as a whole. We know him as composed of distinct parts; and even these parts are created by our methods" (Carrel in *Man The Unknown*).

All discussions of man, his native environment, his natural and legitimate habits, must be conducted in the face of the fact that we have no science of Man, as Carrel often asserts in his book; and we have very little reliable information to guide us through the wilderness and confusion created by the many authors who have sought to advance their opinions, record their views, and register their conclusions as to a subject on which nothing much exists but vague theories and speculations.

Carrel says that modern science has grasped only certain aspects of man. What are they? How shall we know whether these aspects believed to have been grasped are aspects of fact or of fancy? If modern science does not apprehend man as a whole, what parts does it apprehend and what parts does it omit? How shall we know whether the parts apprehended are correctly or incorrectly adjudged?

If we know man as being composed of distinct parts, and have created these parts by our methods, how shall we know whether the parts so created are true to the nature of man, or whether they are the products of the fanciful mind of the Evolutionist? If modern science knows so little about modern man and admits it, there is no reason to believe it knows anything about that man who lived so far back in the remote past that there seems to be little reliable evidence extant to guide us in describing him or any part of him, or his environment or any of his habits.

Men Of Great Stature

We mentioned the giants described by Bharati that lived in the Golden Age, the Silver Age and the Copper Age. Modern Science will meet the report with a smile and pity the ignorance of him who made it.

The Bible should be good authority in this respect, as in it there are frequent references to "giants" being "on the earth in those days, and also after that" (Gen. 6:4). All the people we saw in it (Canaan) are men of great stature. There we saw the giants, the sons of Anak, which come of the giants; and we were as grasshoppers compared to them (Num. 13:32, 33). The Emims dwelt there in times past, a people great and many, and tall, as the Anakims; which also were accounted giants as the Anakims (Deut. 2:10, 11). The Amorites dwelt in the hill country of Canaan. His height was like that of the cedars, and he was strong as the oaks (Amos 2:9). In Deut. 3:11 it is said that only Og, king of Bashan, remained of the remnant of giants. His bedstead was of iron, nine cubits long and four cubits wide, or about 13.5 feet long and six feet wide.

The Valley of Hinnom was known as the valley of the giants (Jos. 15:8). Goliath of Gath was six cubits and a span in height (1 S. 17:4). That would make him approximately ten feet tall, — the height of man in the Copper Age. The staff of Goliath's spear was like a weaver's beam; and his spear's head weighed 600 shekles of iron (1 S: 17: 1).

R. M. Johnson, D. C., said in the Chiropractic Record of July 1916, that in 1830 a living man 18 feet tall was exhibited at Rouen. He added, "'A few years later, near the same city, was found a human skeleton 19 feet long. Three human skeletons unearthed near Palermo measured 21, 30 and 34 feet in length." In the press of July 27th 1930, Ripley reported in his "Believe it or Not," the case of Angoulaffre, the Saracen giant of the 8th century A.D., who was 12 feet tall.

The American Weekly of October 5th 1947, mentioned the case of Jan Van Albert, age 44, who was 9' 3½" tall. The press of February 14th 1936, reported that the skeleton of a gigantic man, with head missing, was unearthed at El Boquin, on the Mico River, in the Chontales district of Nicaragua. The ribs were a yard long and four inches wide. The shin bone was too heavy for one man to carry. In the press of March 26th 1933, Ripley reported the case of Fedor Machnow, Russian giant of Charkow, who was 9' 3" tall. In 1934, a Los Angeles dispatch reported that Dr. F. B. Russell had found in Death Valley, California, mummified remains of a race of men eight to nine feet tall.

Robert Wadlow, of Alton, Illinois, who died in 1941, was 8' 9½" tall, and weighed 490 pounds. The press of December 28th 1941 reported the case of Cliff Thompson of Milwaukee, Wisconsin, who was 8' 6" tall.

Historians of ancient Rome say that the giant Pusio and giantess Secundilla were 10' 3" tall. Pliny wrote of the Arabian giant, Gabbaras, who was 9' 3" tall.

Stature Originally Gigantic

Gobind Behari Lal in the American Weekly of September 17th 1944, stated: "Science has at last proved what it long suspected—that those fairy tale giants really did exist. The time was more than half a million years ago, and they were not occasional freaks, like the circus giant, or Goliath of the Bible, Gigantopithecus, whose early existence was recently reported by Dr. Franz Weidenreich of the New York Museum of Natural History, averaged seven to nine feet tall, weighed 500 to 600 pounds, and had teeth six times the size of modern man's and twice that of a gorilla. It is possible that an oversized Gigantopithecus may have stood 12 feet tall and weighed at least half a ton. The most startling implication is that when the evidence is all in, he may not be the all-time human heavyweight, but merely a middleweight. Gigantopithecus flourished between 500,000 and 800,000 years ago in Java, and man has been shrinking in size ever since. The farther back we go, the bigger we find our ancestors."

The first clue to Gigantopithecus appeared in the shop of an herb doctor in Hong Kong who had three "giant's teeth" for sale. Dr. Weidenreich saw the teeth and was convinced they were human teeth. But this seemed preposterous, as they were twice the size of the biggest ape teeth. In 1941 a jaw-bone was found in Java, unquestionably human, but of such gigantic size that it could have held those "giant" teeth. Between 400,000 and 500,000 years ago there was Meganthropus of Java, still a giant, but weighing only 400 pounds.

Some 200,000 years ago there was Java's Nandoeg Man, with another marked reduction in size. Then came the Wadjah Man of Java, comparable with the Neanderthal Man of Europe. Continuity, periodicity and evolution or development are fundamentals of Manifestation, as we find it. Races of men have inhabited the globe for many millions of years according to the testimony of Ancient Wisdom; and the stature of all things was originally gigantic compared with the period of recorded history. In prehistoric days, giant animals roamed the earth. Some species of the Dinosaurs were 60 to 80 feet long and their weight exceeded 35 tons. These species are said to have been herbivorous only, but had such defective dentition as to make impossible the mastication of food. *Did they eat?* The theory has been advanced by some scholars that man was not originally an eating animal. Some scientists show that what he eats now does not nourish his body.

It is said the cells of man's body are self-existent and self-sustaining, the same as other objects which are composed of atoms, as the body cells are; and that man should have no more need for food than a stone or a star. The body is built of cells, the cells are composed of molecules, the molecule is composed of atoms, and the atom is composed of electrons. Within the molecule in the three states of solid, liquid and gas, everything remains the same, i.e. the proportion of substance to emptiness does not change. Electrons remain equally far from one another inside the atom, and revolve in their orbits in the same way in all states of cohesion of the molecule. Changes in density of matter, i.e. transition from solid to liquid and gaseous states, do not affect the electrons.

Man's Body Resembles The Planetary Bodies

The area inside the molecule in the living cell is exactly analogous to the vast space in which move the celestial bodies, which space scientists assume to be void and empty. Radio proves that theory is erroneous. Electrons move in their orbits in the atom in the cells of the human body just as planets move in their orbits in the solar systems of the Universe. The power of movement in all these instances comes not from food. There is one universal law that governs all bodies.

The electrons in the atoms in the cells of man's body are the same celestial bodies as the planets of the sky. Even their velocity is the same. This now established fact shows how completely man's body resembles the planetary bodies. The Ancient Masters were right when they said, "As above, so below." Man is microcosm and the Universe is macrocosm.

If the planetary bodies depend not on food for their origin and maintenance, neither should the human body. One universal law governs all bodies. *Why does it seem that man dies of starvation when deprived of food?* Because his body has adapted to that kind of stimulation.

Somatic death results from the shock of the body being deprived of a certain kind of stimulation. It is the loss of that stimulation which activates the vital machinery of the body, and to which stimulation of the vital machinery is geared, and adapted, and accustomed by reason of ages of practice. As man adopted new habits through the ages, it was necessary for the body machinery to adapt itself to the new practices or perish. The adjustment occurred slowly, gradually, the body yielding to the new habits through the Law of Vital Accommodation only when the primary reaction was disregarded and the practice continued.

Each phase of the adaptive process can occur only at the expense of a depression of all the vital functions, which must be injurious if long-continued or often repeated. Man never turns back. He was proud of what he terms "progress" as he stupidly advanced from perfection to imperfection and greater imperfection. He believes that each new practice is an improvement, provided the effect of it does not kill him instantly. So he continues on his descending course, while his health steadily declines and his life-span constantly decreases. The basic causes are never suspected.

Misleading Reports

Deceptive propaganda is often put out to show man's life span is increasing. People believe these reports. They are false and calculated to deceive. It is the "life expectancy" that has increased, and not the life-span. The "life expectancy" has increased because more babies are living than they did fifty years ago. In the press of July 30th 1929, Professor C. H. Forsyth of Dart-mouth College said: "The expectation of life from the age of 45 on, is the lowest of which we have any record. Far lower than it was only forty years ago — and it is still falling, not rising."

Carrel wrote, *"A man of 45 has no more chance now of dying at the age of 80 than in the last century. . . . Not ever one day has been added to the span of human life"* (Man The Unknown, P. 178).

Man has died suddenly upon receiving bad news because his body could not take the shock. Many a man has died because the body could not take the shock of being deprived of a certain poison, as opium or morphine, which his constant use had forced his body to adapt itself (to that stimulant so completely), that its use became a physiological necessity in the vital economy of the organism. The same is true of food.

Doctors know that it is dangerous to stop any practice too abruptly. No doctor will claim the body needs opium. Yet man sometimes does when deprived of it. It is a matter of common knowledge that the drug addict dies by inches from the effect of the poison which his body craves, and that it is dangerous to deprive him of it suddenly.

The more poisonous a substance is, the longer it takes the body to make the adjustment, the greater the adjustment from the perfect point, and the greater the shock when the body is deprived of that substance.

The greater the shock, the more dangerous the shock. The drug addict almost goes crazy when deprived of his poison. The nerves of the smoker practically go out of control when the last cigarette is gone.

Body Craves Food As It Does Poison

That is a typical example of the operation of the law of Vital Adjustment. Due to the established fact that the body will adapt itself to the point of craving to the death of a destructive poison, it is not difficult to understand that, by countless ages of practice, the body has become so firmly and completely habituated to the stimulating effect of food, that it will gradually sink down in death when deprived of that kind of stimulation. We must remember that it is stimulation and not nutrition in this case. If sudden death results from the physiological shock of depriving the body of a certain poison, it should not be mysterious that death by steps and stages occurs from the shock of depriving the body of another form of stimulation called food. It is well to consider that both eating and drinking are voluntary and controlled practices. Not only are these under man's conscious control, but he can go without food for weeks and without water for days — but to stop the breathing is to stop the living. This evidence proves beyond question the paramount importance of air to the body. Furthermore, the function of respiration is an automatic, involuntary process, and so far beyond man's conscious control, that he breathes when unconscious in sleep, or from injury; even better and deeper, more regularly and rhythmically, than when conscious and awake. Furthermore, polluted air disturbs the sleeper more than it does the worker. Quite often the cause of insomnia is bad air.

It is impossible for one to commit suicide by holding the breath. As soon as consciousness is lost, breathing automatically begins again. It is neither reasonable nor practical that man's body is so constituted that its existence on earth depends on the exercise of voluntary and controlled processes, such as eating and drinking are. The breathing function is not only automatic and involuntary, but the primary function of the living organism. All other functions are secondary and designed to keep the body fit to perform Respiration.

The lungs are definitely constructed and constituted for their work. They are by far the largest organs in the body, filling the thorax from the clavicle to the floating ribs and from the sternum to the spine. In comparison, the stomach is an insignificant organ, being only an enlargement of the alimentary canal that extends from mouth to anus.

Man Eats To Die

In the 19th century a group of eminent doctors, composed of Hufeland, Rowbotham, Raymond, Rayer, Gumbler, Month, Freille, Bailey, Winckler, Easton, Evans and others, working independently, made a special study of the causes of Old Age and death from so-called natural causes. Without exception they all arrived at the same conclusion by finding that the cause is — "We eat to live, and we eat to die" (Densmore, P. 284). If we eat to live, how can we eat to die? If we eat to die, how can we eat to live? These doctors found that we grow decrepit and go to the grave because of what we eat, and that appears as definite and irrefutable evidence to prove that eating is not natural.

All authorities who have investigated the matter, agree that as man descended from the Breatharian State, he first ate fruit. Then came the time when he began to fish and hunt, and he added flesh and fish to his diet. He next became a "husbandman" (Gen. 9:20) and began to till the soil, and increased his diet with the addition of vegetables and cereals.

These authorities place fruits and nuts first in their fitness as food; animal products are placed second, vegetables third, and last and worst, pulses and cereals. The latter, due to their excess of earthy minerals, are most suited of all foods to induce ossification of the tissues and joints, thickening and hardening of the muscles and arteries, and consequent and inevitable premature Old Age — that decrepitude and imbecility almost universal, but wrongly reckoned as a necessary condition of senility.

Each step man took he considered progress — as he still does. It was progress in the wrong direction — as it still is. But he never turns back. As he took the first step of this "progress," his health began to decline and his life-span to decrease — but the cause was a mystery to him.

When the cause was definitely discovered, it was too late to do much about it. Man's body had suffered such serious degeneration and had so completely adapted itself to the new regime, that he refused to resist his craving for it, finding it easier to sicken and sink into an early grave than to fight against the bad habits that had enslaved him as he "progressed." In this age he is still progressing, still adopting new habits, while holding hard to the old ones, and is steadily sinking both mentally and physically — while "medical science" wages its "bitter battle against disease," as was boastingly declared in the press of February 15th 1925 by Dr. Morris Fishbein, late editor of the Journal of the A. M. A.

Lesson No. 13
Chronic Auto-Intoxication

When Dr. Evans in 1879 wrote in his "How to Prolong Life," that, *"We eat to live, and we eat to die."*

He made a statement that inspired us to search and see whether we could find an answer to this riddle. We found that a vegetarian diet causes cretaceous degeneration, atheroma, arteriosclerosis, hardening of the blood-vessels, ossification of the tissues and joints, and premature decrepitude or aging. Furthermore, alcohol is made of grain; and the production of alcohol is in constant operation in the bodies of those who eat grains and cereal products.

These vegetarians are always in an auto-intoxicated state, to which their bodies have become so completely adjusted, that when the toxic effect begins to fade, an uncomfortable feeling appears in the sensations of hunger, weakness, nervousness, headache, etc.

The doctors warn these people to be regular in their eating habits. That keeps the body in its chronically intoxicated state, the condition to which it has become adapted. So they feel better and regard the doctor as a wise man. The condition is similar to that of the chronic drunkard sobering up. He is weak, nervous, has headache. These symptoms simply show that the body is calling for more liquor to keep it in the semi-paralyzed, auto-intoxicated condition to which it has become adapted.

The more tobacco or liquor one uses, the more the body craves it. The more food one eats the more the body craves it. The more flesh and cereals one eats, the more the body craves them. The basic cause is the same. In each case the craving shows that the body has adapted itself to these damaging poisons. The flesh-eating man is in a steady state of mild irritation, or auto-intoxication, rising from the excessive stimulation caused by the decaying flesh food in his stomach, bowels and blood.

Just as in the case of the flesh-eater, the vegetarian has eaten his diet from childhood, and the condition of auto-intoxication is deep-seated and chronic. The victim's body and organs are adjusted to it.

Another reason why man feels hungry is because flesh protein starts to putrefy as soon as eaten, and the sensation of hunger results from the toxic condition that rises. He eats bread, and this starchy product ferments in the intestines, producing alcohol and acids that temporarily dull the putrefactive bacteria that work on the protein remnants of the flesh formerly eaten, thus suppressing the effect of auto-intoxication, and the fatigue produced by the auto-intoxication temporarily subsides.

Body Tries To Maintain Balance

After a meal of bread, meat and potatoes one feels strong. "Food gives you strength," say some people. What actually occurs is only a stimulation of the tissues and intestines that increase peristaltic action, and their contents are moved along, while the carbohydrates suppress the protein putrefaction. Man eats a sandwich of bread and flesh and feels stronger because the bread counteracts the rotting of the flesh formerly eaten. A few hours later the last flesh eaten starts to rot, and the sensation of new hunger causes him to eat another sandwich, which temporarily counteracts the toxic weakness mistaken for hunger.

That is the vicious circle. Hunger is usually a call for food to counteract the effects of toxins of the food previously eaten and to stimulate the intestines to move along their contents. Ehret reported cases of religious devotees who had fasted for decades. He held that the purer the body, the less it craves food. When the body is perfectly free of toxins and in normal condition, the need for food practically vanishes. When there are toxins in the body, acids form in the fluids, creating a need for alkalies in the form of fruits and green vegetables, which supply the body with activity in cell function by stimulation. Man realizes not how filthy his body is internally until he takes a fast and the body begins to purge itself of accumulated filth. The body of the average man is filled with accumulations of filth where the body has adjusted itself and compensated for errors and indulgences. When we stop the errors that age and disfigure us, these accumulations start to dissolve and depart via the eliminative channels — the skin, lungs and bladder.

When headache lassitude, vomiting and skin eruptions appear as the purging process continues, the details of which we know nothing, most people get frightened and they are advised to "stop this nonsense of not eating." As soon as they begin to eat, that stops the purification process, and they feel better as the eliminative processes cease. They think eating is necessary and that fasting is dangerous, as medical art teaches. Did they not feel better as soon as they began to eat? They feel better because the eliminative process was halted. One feels bad as the body begins cleaning out obscure corners and this foul mess begins to leave the body. This filth within the body is in the form of fat, lumps, blind boils, tumors, cancers, growths of all sorts, silent within the body.

We blame much on the weather, when it is the filth being eliminated because the sudden changes in temperature have made the body forces active, aroused and awakened them, brought them into action, perhaps by making an opening in the heavy coating of corroding filth around the poles of the cells which obstructed contact with the Life Principle. Feasting fills the body with filth and disorders, while fasting enables the body to cleanse itself. Dr. Pearson reported that he removed a tumor from a patient of over twenty years starting with a fast of 30 days. Dr. Eli G. Jones reported the removal of cancers by fasting. Simply remove the Cause and you have the Cure.

Vitality Increases

Sometimes fasting patients begin to vomit about the 20th or 25th day, and may vomit intermittently for two or three days. Then they get very weak. But after the vomiting rids the body of filth stored in it for twenty or more years, they are amazed by the increase of strength.

This filth corrodes the poles of the cells and interferes with the reception of Life Force. As the filth is removed and eliminated, better contact is established between the cells and Life Force and the patient's vitality immediately increases. This fact is known only to doctors who have fasted patients, while medical art will claim the statement is false. No orthodox doctor dares to make a test of it.

On account of ages of habitual eating, we cannot trust our apparent instinct and craving for food. We should combat the craving until the habit is conquered, as with all habitual errors, indulgences and abuses.

A man living in good, fresh air can break more of Nature's Laws and reach a ripe old age than one living in foul, stagnant air. Evidence of this appears daily in the press of people who live to be 100 or more, who have broken many health laws — smoking, drinking, gluttonizing, sex abuse. In such cases we find that the persons have not housed themselves, have not lived on a main highway, nor worked in garages, filling stations, factories, stores, offices.

Eat Little And Live Long

The theory that food is so essential to sustain the body receives a heavy blow whenever investigated. As little food is eaten the result is better health and longer life. The 1938 edition of the Encyclopedia Americana reported that Bulgaria was the leading land of Centenarians in the civilized world, and they subsisted on a frugal fare. The account said: "People (in Bulgaria) said to be 105 to 125 years of age are not uncommon.... Over 70 percent of the people are engaged in agriculture.... Fruits and vegetables are raised in abundance.... Wine is plentiful and cheap" (Volume 5, P. 1).

A commission of Bulgarian doctors visited a large number of these old people in 1927, and found they lived on a simple diet of fruits, vegetables, sour milk and buttermilk. They were lean to the point of underweight according to medical standards. Only one of the group was found to weigh as much as 168 pounds, the majority weighing between 122 and 130 pounds. Bulgaria had 58 centenarians per 100,000 of population, while the U.S.A., the land of plenty and gluttony, had but 4 per 100,000 population, with the number of centenarians steadily falling.

The press in 1938 reported the case of Kakudo Yamashita, a cemetery caretaker in Tokyo, who lived for 35 years on nothing but uncooked free leaves, grass and weeds. He was then 80 and reported in good health. K. L. Coe, writing in *Correct Eating* for March 1931, reported that Dr. Robert McCarrison, of the British Army Medical Service in India, found in a colony in the Himalayan region that the natives were so old it was hard to believe their records were correct, yet he was unable to detect any error in their method of keeping time.

Coe said: "Ages well beyond 250 years were common. Men of well attested age, up to 150 years, were recently married and raising children. Men said to be well over 200 were working in the fields with younger men, doing as much work, and looking so much like the younger men, that McCarrison was not able to distinguish the old from the young. These people live long and are free of all disorders because they spend their time in the open air and subsist entirely on simple, natural foods, with a small amount of milk and cheese."

An account appeared in a certain magazine in 1950 to the effect that one Oswald Beard, an English veteran of World War I, was wounded in the stomach, and for the last ten years had been able to take nothing but tea "spiked with plenty of cream and sugar." No dietetic expert will agree that such a diet will support the body.

Lesson No. 14
Body Needs

"Conquer this foul monster, Desire, most difficult to seize, and yet possible of mastery by the Real Self; then bind him fast forevermore, thy slave instead of thy master" (Message of the Master, P. 44). "To want nothing is divine. To want the least possible brings one nearer to divine perfection. The less physical man becomes through the conquest of his Desire, the less he needs. The less man needs, the nearer he becomes like gods, who use nothing and are immortal" (Socrates).

Dietetic experts are like other experts. They think their theories are right when they are most always wrong. They contend that man must eat this for protein, that for carbohydrates, something else for vitamins and minerals, and so on. They forget the fact that the cow eats only grass and green leaves and lives in good health. One test case is sufficient to show the experts are wrong: In its "Believe it or Not," in the press of January 16th 1934, Robert Ripley, who said he had evidence to prove the truth of everything he published, reported the case of Jekisiel Laib, of Grodno, Poland, who fasted six days a week for thirty years. Each Saturday he ate bread and water, and his health was good. Laib declared that eating is an acquired habit and not natural, as are chewing and smoking tobacco, and he is determined to prove it. Modern man subsists on a diet of demineralized substances that form dangerous acids in his body and rob it of minerals. How does he keep going on such a diet? How did Laib keep going on a diet of bread and water? Because man has always been and still is a Breatharian. He gets his real nourishment from cosmic rays, from the air — cosmic food.

The food man eats, like the tobacco he smokes, and the liquor he drinks, are only stimulants and indulgences that have little to do with the process of nutrition, but rather interfere with it. He who eats demineralized food and gets sick, gets well when put on a fast and eats nothing. Every man who eats is an example of one who lives without getting his nourishment from food. What he eats robs his body of its minerals, due to the acid-forming effects.

Minerals From Cosmic Rays

We eat some spinach. Does it give the body iron as dietetic experts claim? What actually happens is this: Practically all foods we eat form acid toxins in the cells, as do proteins, fats, and carbohydrates. These acid toxins interfere with the cell's function to assimilate minerals from cosmic rays. Dr. Crile asserted that death occurs at the moment when positive acidity is established in the body. He showed that — "All inhalation anesthetics cause a progressively increased hydrogenation concentration, with death occurring at the moment when a positive acidity is established." The acid-alkali balance of the body has a vital significance. When the alkalinity of the blood falls, the animal dies.

This balance between the nucleus and the cytoplasm of the cells (the electric potential) is essential to animation, and supplies the vitality of the living process itself. Its reduction to zero produces death. The body cell is bi-polar, a miniature battery, whose nucleus is the positive pole and the cytoplasm is the negative. The cytoplasm of the cell surrounds the nucleus and is the negative (alkaline) element. The nucleus is the positive (acid) element. The cytoplasm is a colloidal solution of alkaline minerals. This makes the cell a bi-polar mechanism.

The essential, characteristics of vitality manifested in the cells, as assimilation, growth and generation, depend not upon food, but upon the presence of an electric potential produced by oxidation. The various functions of the organs are due to a variation of the potentials.

The foods we eat form acids, and the acids reduce the alkalinity of the cytoplasm; thus, reducing the cell's electric potential. This means reduced capacity of the cell to respond to cosmic rays, which means mineral deficiency and physical weakness.

We attempt to overcome the condition by eating mineral-rich, alkaline vegetables. They introduce alkalies into the blood that temporarily neutralize the acid toxins, causing the cytoplasm to become more alkaline, and the cell's electric potential increases. It regains its capacity to assimilate minerals from cosmic rays. Then the dietetic expert steps in and gives the spinach the credit, asserting that the minerals of the spinach were transformed into cell minerals. That is erroneous and just more dietetic nonsense. The minerals of the spinach are largely eliminated through the bladder, in the form of neutral salts, by combination with acid toxins. Practically all the minerals of foods are eliminated in the urine in the form of neutral salts.

Sensation Of Hunger

Each kind of food induces its own type of stimulation, producing a hunger sensation — one may test it and see. Eat bananas until no more of that type of stimulation is desirable, and then change to pie or cake and a new hunger sensation appear. That is the reason why one eats with a "coming appetite." A man, not hungry is persuaded to eat. As he eats a feeling of hunger develops and he eats a big meal. This course can be continued until the stomach is distended to a discomfortable degree.

Such a state of distress may result that vomiting becomes necessary to relieve the outraged stomach. Then the glutton takes some dope to settle his stomach, and that adds insult to injury. Maybe a doctor is called, who administers some of his scientific poisons to make the unruly stomach behave and submit without complaint of abuse.

It requires the concoctions of "medical science" to whip the body's organs into line and make them stop rebelling against maltreatment. The organs should know better — since they do not, "medical science" teaches them a lesson, making them more docile and less unruly.

Eating Is A Vicious Circle

We eat acid foods and then must eat alkaline foods to balance them. If we ate no acid foods in the first place, we should have no need of alkaline foods. Once our body is clean and pure, we should have no need to eat. But eating is necessary until vital adjustment is established so the body cells no longer require that kind of stimulation furnished by food.

Biochemists show that a plant possesses a laboratory that transforms invisible substance of the air into cell material. The cells of the leaf split the water molecule of the air into oxygen and hydrogen, rejecting the oxygen atoms and appropriating the hydrogen atoms for replacement cells. The experiments of Richards showed that in the plant's laboratory the disintegration of acids through the division of the acid compounds is not a digestive, but a respiratory process, and results from the alternate oxidation and de-oxidation of the plant tissues through the action of cosmic radiation. He held that the same principle applies to man.

Biochemists assert that the function of the intestinal tract is to serve as a laboratory for condensing and combining nascent hydrogen gas with nitrogen gas entering the lungs as air. They regard the lungs, not the stomach, as the chief organs of digestion, with the intestinal tract serving only to carry on its primary function of condensing, combining and eliminating. Biologists hold that oxygen supports the cells, and that nitrogen acts as a tissue builder and vitalizer. Cosmic nitrogen appears in the muscles and fibrous tissues, and is said to be the first of all the elements to leave the body when it is dead. Physiologists declare that without hydrogen, creative activity in the body would cease. There would be no salivation, no perspiration and no elimination.

It appears that hydrogen also soothes the nerves, regulates body temperature, moistens the lung surface, carries off toxins, cools the tissues and retards inflammation. Without hydrogen, the nerves and tissues would stiffen, harden, and decay. Physical science declares that the body is animated, vitalized, nourished and maintained by the material food eaten, digested and assimilated by the cells through a direct chemical process. No orthodox doctor dares question that scientific theory, and no scientist ever lived who could analyze and explain the process in an intelligent and logical manner — because that alleged process is definitely imaginary. The body is built and composed of cells. The cells are composed of molecules, the molecules of atoms, and the atoms of electrons, which are found to be whirling centers of force in the ether. We may reverse the process in our mind and see those whirling centers of force transforming into visible substances, forms and entities by a natural process of condensation resulting from a retardation of the vibrations.

It is common knowledge that invisible gases condense to form ice. It should be as easy to understand that, under the same law, these invisible gases condense to form blood, flesh, bones, and the entire body — according to a pre-existent pattern, which is that "building of God, an house not made with hands, eternal in the heavens" (2 Cor. 5:1).

Atomic Energy

Splitting the atom proves that a mighty force is inherent in cosmic substance which condenses and forms visible bodies. Its modus operandi never alters, and it acts constantly in man, in his cells, operating; simultaneously on all planes of being — yet appearing to manifest itself only in material forms. That Cosmic Force, controlled by Infinite Intelligence, possesses the property of activating and stimulating atoms to assume various relationships under the Law of Polarity, and galvanizes the power latent in dormant cells into a higher state of activity. We see in radio-activity that matter is transformed into radiations. Lakhovsky said that the stars are radioactive, and their various minerals, in the form of electrons, constantly pass out into space in the form of cosmic rays. When cosmic rays strike another part of the cosmos, they materialize into the identical minerals from which they arose.

The chromosomes of our body cells are constantly exposed to a bombardment of cosmic rays. The chromosomes pick up these rays, which are transformed into (1) vital electric currents, (2) nerve and brain electricity, and (3) electrons of various Minerals. The cells of the body are sustained by these minerals, and the cells manufacture all their constituents from cosmic rays.

The cow subsists on a diet of grass. We examine the grass and find no fat in it. Whence comes the rich butter fat in the cow's milk? It is manufactured in her body from cosmic rays. Most people live on a demineralized diet that drains many elements from their bodies. Yet these people seem to be healthy and energetic. Whence come their minerals? From cosmic rays. If they depended on food for their minerals, they would soon be physical wrecks because of their demineralized diet. In spite of their mineral losses, they are constantly receiving minerals from cosmic rays, and not from the demineralized food they eat. If man's body were not sustained by cosmic rays, he would shrivel up and disappear in short order. He is still a Breatharian.

Lesson No. 15
Eating Poisons

"The ways of the kitchen and dining room are the ways of disease and death, ways whose ends are prisons, asylums, scaffolds, to a far larger extent than is dreamed of by the mothers and fathers of the land." — The Fasting Cure by Edward H. Dewey, M.D.

When man eats, he not only makes of his body a distillery that produces alcohol, but a refinery that consumes various poisons. Civilized man constantly eats poisons and knows it not. His foods, such as potatoes, tomatoes, lettuce, spinach, onions, carrots, beets, etc., are bad enough, but to the poisons these contain are added the deadly insecticides used to protect these crops from the bugs. The vegetarian thinks his diet is superior to that of the carnivorian, and knows nothing about the actual poisons his vegetables contain.

1. The common potato, called white or irish potato, is of the nightshade family. Its botanical name is solanum tuberosum and its native country is the Andean region of South America. This tuber contains two narcotic alkaloids, one of which is solanine, and in some cases it causes "potato poisoning" that requires hospital care. Few animals will eat potato vines because of their poisonous properties. Fowls will not eat potato bugs because they are so poisonous. Potatoes are also used to make alcohol.

2. The tomato belongs to the nightshade family. It is a native of South America and was taken to Europe in the 16th century. Tomato vines are as poisonous as potato vines, and few animals will eat them.

3. The onion contains a soporific substance and an irritating oil, which makes the eyes water and the genital mucous membrane do the same. Some authors assert that it acts as a powerful aphrodisiac, besides irritating the kidneys and bladder. Garlic does the same, only to a greater degree.

4. Lettuce got its name from its milky juice, lactis, milk. Lactuca serriola is believed to be the wild variety from which the cultivated kinds were derived. It contains a soporific, harmful narcotic alkaloid termed lactucarium, and is sometimes used as a substitute for opium.

5. Asparagus, celery, cabbage, onions and turnips contain saltpetre. Cabbage and turnips also contain arsenic. Beets, eggplant, spinach, swiss chard, rhubarb, all contain certain poisons, and the poisons are the properties that force the body to adjust itself to the point where it craves these stimulating substances.

Potatoes, lettuce and practically all so-called vegetables dull the brain and produce enervation. One may speak of solanine psychosis or potato psychosis or lettuce psychosis as a mental disorder caused by eating these substances as one speaks of alcoholic and opium psychosis.

Dangerous Narcotic From Juice Of Poppy

The poisonous properties of opium are well-known. A full dose is intoxicating and exhilarating, but its effects are damaging and fatal. It is one of the most energetic narcotics, and is made of vegetable juice — the juice of the white poppy. The tobacco plant is a vegetable, a herb. It has become one of the great mediums of commerce and is now widely used all over the world. Tobacco, in medicine, is a powerful stimulant, emetic, and purgative. These various effects appear as the reaction of the body's protective force to poison. As the body is thus poisoned it suffers permanent damage, resulting in a decrease of the health and life-span.

In spite of the poisonous properties of the vegetables mentioned, they are freely used by millions of people, who also use as food other vegetables containing poisonous properties, not so pronounced as those of tobacco and poppy plants, but so damaging that it has been necessary for the body to adjust itself to their use and effect. Such adjustment can occur only as a result of a decrease in the body's vitality.

The Breatharian contends that eating is an acquired habit, as are the chewing and smoking of tobacco, and that it has been necessary for the body to adjust itself to all substances man eats, just as the body, under our observation, must adjust itself to the use of tobacco, opium, morphine, medicines, drugs and all other poisonous substances. The vegetarian's body is stimulated and saturated with poisons of the substances he eats; and the vegetables one craves most are those that poison one most. The vegetable poisons are usually mild in character and slow in action, but their damaging effects are cumulative, increasing with the years, and are responsible for many disorders, the basic causes of which are never suspected. The vegetarian goes through life in that weakened condition, with his vitality much below par, but he knows it not because he never knew anything else. He has no reason to believe his vitality and health are not what they should be. He cannot miss what he never had. He is dying by inches all the days of his life, while believing that he is following a high standard of living.

Mice Unable To Live On Human Diet

In the press of March 8th 1938, headlines appeared reading, 'Man's diet fatal to mice in test by pathologist." The account said: "Washington, D. C., March 8th (AP) - Wanted, a strain of mice that can stand up under a human diet of cocktails, highly seasoned foods, hot foods and drinks. Dr. Maud Slye, Chicago pathologist, said today that she has been looking for years for such a strain. She added: 'The diet of man containing these three properties kills any mouse that I have ever tried on it'."

This is the conventional diet of the man of today, and people are constantly urged to eat freely of it to build up their "resistance to disease." The body-adjustment-powers of a mouse are less efficient than those of man; so man is able to eat, and not drop dead on the spot, what the mouse cannot take. On that diet man dies by inches, and the basic cause of his demise is never suspected. In the July 1951 issue of "American Magazine," appeared an article by U. S. Congressman James J. Delaney, "Chairman of the nonpartisan House Select Committee to investigate the Use of Chemicals in Food Products," in which he relates some findings of the committee that are startling in showing how food products are chemicalized with various poisons. We quote: "Not long ago a frozen-food packer was told that his new shipment of peaches would stay bright and fresh-looking if he added a touch of thiourea. He tried it. The chemical worked a miracle of freshness and coloring. The shipment went out. Another frozen-peach firm did the same thing. Before shipping out its product, it invited the local Food and Drug Administration inspectors to test the food. Samples were fed to experimental rats. Within a few hours they all died. "Several years ago a salt substitute was put on the market for use by people on a low-salt diet. It contained lithium chloride, a chemical whose effect had been only superficially tested. Three persons died before the 'salt' could be withdrawn from the market. These instances point out a blunt fact: Our food supply is being doctored by hundreds of new chemicals whose safety has not yet been established." — P. 19.

There is no use covering more space with quotations from the article. It may all be summed up by saying that people are being poisoned daily not only by what they eat, but by the poisonous chemicals added to foods to improve their appearance, their flavor and their keeping qualities.

Under the Law of Adaptation, the body is able to adjust its physiological processes to a surprising degree in favor of any substance, no matter how poisonous; and it appears paradoxical that the more poisonous the substance, the more the body craves it and the greater the shock when the body is deprived of it. By degrees the body will adjust itself to tolerate increasing doses of a poison until the amount taken at one time may be so large that it had killed the man quickly had he taken that amount at first. In the matter of food the same rule applies. It seems that the most harmful substances are those which the body craves most, due to the greater adjustment the body must make in favor of these poisons.

The more deadly the poison, the greater the degree of the body's adaptation; and the greater the adaptation, the greater the shock when that particular poison is withheld. The substances that are the hardest for man to give up are those that cause the body to make the greatest adjustment to endure them. These as a rule are the substances that contain the most or deadliest poisons. As we ascend from vegetables to fruits, we find that the fruitarian takes into his body less poison than does the vegetarian, assuming in both cases that the substances have not been sprayed with insecticides. Man does not entirely free his body from the ingestion of poisons until he becomes a liquidarian and drinks only pure snow or rain water in regions where the air is not polluted by civilization.

If the air is polluted, the snow and rain falling through that air is polluted to a certain extent. Poisons still enter the body of the

Breatharian when they are in the air. The only air free of poisons and fit to inhale is that found miles from the centers of civilization.

Evolution And Devolution

Some ask why man has teeth and alimentary canal if he was originally constituted to subsist on air and cosmic rays only. The body contains many dormant, rudimentary organs which once were developed and useful, as God makes nothing in vain. The teeth complete the mouth and aid man in talking. Certain developed organs in the body of the Eater will shrink and become dormant when man again becomes a Breather. There is a prevision and a provision within the living organism, by which it may anticipate future conditions, and may rise superior even to heredity as well as environment, until it meets and masters the conditions of a progressive or an established achievement. The prevision is the power to realize the ultimate effect of the unnatural use of any substance or thing, and guard against this effect by vigorous reaction, yielding to the inimical influence through the Law of Vital Adjustment only when the primary reaction is disregarded. We must never forget that such adaptation can occur only at the expense of a depression of all the vital functions, which must be injurious if long-continued or often repeated.

On account of its theory of Evolution, physical science must see in man only an object that has existed always in his present shape and form not only, but has actually improved beyond the ape-stage where he once was. Consequently, his habits have made only minor changes in order to keep pace with his improved organism and his higher plane of existence. It is this theory which physical science must protect by opposing everything that tends to explode it.

Contrary to the scientific theory of Evolution, there is not a scrap of evidence to show that any creature has ever improved when left to its own resources. On the other hand, the rule is that all things degenerate unless given proper aid directed by proper intelligence. The human body was perfectly constituted and designed in the beginning to meet and master all conditions to which it would be subjected in physical existence on earth. When man was a Breatharian, his body was equipped, in a rudimentary degree, with all the parts that would later be needed, in a functional degree, to prevent him from becoming extinct as a species, regardless of what habits he might adopt or the condition of the environment by which he might be surrounded. Evidence of this fact appears in the case of modern woman, whose clitoris is nothing more nor less than a rudimentary penis, as Dr. G. R. Clements showed in his work, *Science of Regeneration*. Why does she have that rudimentary organ when it is of no use to her? Man asked himself the same question when he was a Breatharian. He wanted to know why he had the rudimentary parts and organs which were of no use to him until he descended to Gluttarianism. When his gradual change of habits made it necessary for his continued existence, that the rudimentary parts and organs develop and become functional to save the race from extinction, they developed to a functional degree, and now we think man was always as he is at present. Much light on this appears in Darwin's works. He refers to the law of variation and says: "All organisms exposed to new and changed conditions, vary. Accordingly, there is no case on record of a variable organism ceasing to vary, when subjected to the Law of Variation. The oldest living forms known are still capable of further modification." Variation of Species. P. 5. Kirby and Spence describe in their work under the Law of Variation how bees will change in constitution in one generation. They write: "If bees are deprived of their Queen and are supplied with comb containing young worker-brood only, by having a royal cell erected for their habitation, and being fed with royal jelly for not more than two days, when they emerge from the pupa state they will come forth complete queens, with their forms, instincts, and powers of generation entirely different. But had they remained in the cells that they originally inhabited, they would have come forth as workers." — *Science of Regeneration*, Lesson 48, Chapter 148.

Dormant Organs Ready When Needed

Man's body contains the vestigial remains of organs that once served a purpose. As he descended to lower stages, the organs that functioned on the higher plane were no longer needed, and withered and grew rudimentary. Modern science considers them left-over appendages of the ape stage. Some doctors say we know not whether they are coming or going. They are doing both — "coming" when needed and "going" when not. Dormant organs develop and become functional when needed. If the body contains dormant organs that were developed and functional in the Breatharian Age, they will revive and return to their former state when the demand for their use is supplied.

We know little about man now and less of what he was in the beginning. We know nothing about a normal man if the term means original man. Variations from the original pattern are conditions of Devolution and not Evolution. Physical Science says man is ascending. The facts show the reverse. Nothing left to its own resources ever improves in structure and function. The course is always downward. Otherwise the lowest man, without effort, would become the highest by reason of evolution. We never rise by drifting with the stream. If we did, there would be no incentive for intelligent and diligent labor. Man follows the line of least resistance. Ascension requires work while descension is to drift with the tide.

Lesson No. 16
Vegetarianism Is Bad

"We eat to live, and we eat to die." — Charles W. De Lacy Evans, M.R.C.S.E., late surgeon to St. Saviour's Hospital, London, in his work "How To Prolong Life," P. 28. A puzzling paradox. If we eat to live, how can we eat to die? If we eat to die, how can we eat to live? That's the mystery we are now engaged and considering in this work.

The Vegetarian thinks his diet puts him a big step ahead of the flesh eater; but the records show that the average health and the average life-span of the average Vegetarian are not above that of the average Carnivorian. In his reference to Vegetarians, Dr. Evans wrote: "Cereals and farinaceous foods form the basis of the diet of the so-called 'vegetarians,' who are guided by no direct principle; except that they believe it is wrong to eat animal food. For this reason vegetarians have no better health and live no longer than those around them" (Densmore, P. 303). We shall be guided by a direct principle in this discussion, and learn the reason why vegetarianism is bad, and why many continue to consume flesh in spite of the long campaign waged against the practice.

It is well to remember that the less vital the body is, the more poison it will endure without discomfortable reactions. When danger threatens and the body is too weak to defend itself, that man is in a serious state. Poison will not affect a dead man, and it affects a half-dead man less than it does a more vital man. An example: As the killing cold dormantizes the peripheral nerves of the body, they become unable to convey danger signals to the brain, and the freezing man no longer notices the cold. In fact, he begins to feel warm, and then he becomes unconscious, feels nothing, and freezes to death. Most adults know that tobacco and liquor are bad for the body. They are nerve and brain poisons. The use of these lowers the vitality, and the body, due to a reduction of its powers, adjusts itself physiologically to their use.

Body Vitality Reduced

The same principle applies to eating. When the Breatharian ate his first food, it affected his nerves adversely, just as the first smoke adversely affects the youth. But the youth persists until the poisonous tobacco dulls his nerves and lowers his vitality; and thus his body adjusts itself to the use of tobacco not only, but in time comes to crave the poisonous weed. So the Breatharian continues to eat in spite of the ill effect of it, with the result that the body's vitality was reduced because the brain and nerves were dulled. His body, thus weakened, adjusted itself to the unnatural, enervating practice of eating, and he began to crave food. And it came to pass that as the substances the Breatharian ate weakened his body, man actually eats to die just as he smokes to die.

We must not miss the fact that he who eats has never had the amount of vitality that man's body originally possessed as a Breatharian. We refer not to physical strength, but to that physiological vitality which carries man on through the years and keeps him as nimble at ninety as he was at thirty. For the turning of the earth on its axis has no affect on the body. We must understand that the process of vital adjustment necessary for the body to tolerate a substance it is not made to receive internally, is actually one of vital reduction. The enervated body will endure without protest a dangerous enemy which the vital body wars against and strives to subdue and eliminate. The symptoms of that struggle are termed disease by the doctors. It took time to modify and weaken the body so that it would not only tolerate, but crave the poisonous plants and herbs, termed vegetables, which constitute the diet of the vegetarian.

Most Vegetables Are Not Natural

Most of the vegetables man eats are not found growing wild in Nature. The fields in which they grow are man-made. God made no mistake when he covered the earth with trees. The science of forestry shows that deforestation lays the land open and naked to the fury of the wind, the scorching rays of the sun, and to erosion and destructive floods. Authorities on the subject assert that many waste regions and desserts were once teeming with fertility and foliage; and the existing sterility of these places is the work of man in the destruction of forests.

Man's burdens rise from his foolish efforts to modernize the Primitive and artificialize the Natural. Striking examples of the catastrophe that results from this work are the deserted Dust Bowls of the earth. Fields are artificial; annual crops are artificial; their cultivation is artificial. Fields are the work of man in his effort to "improve" on Nature. The cultivation of these fields is as artificial as the fields themselves. The annual crops grown in these fields are as artificial as their cultivation. All forms of artificializations disturb the equilibrium of Nature, and the results are always disastrous, says Professor John C Gifford, University of Miami, Florida, who wrote: "Fields are sun-baked in dry weather, and muddy and eroded in the rainy season, where cultivation is entirely artificial, and where the equilibrium of Nature has been completely upset by the work of man. All these cultivated things would die and disappear without the pampering hand of man" (Tropical Subsistence Homstead, P. 95). The Vegetarian has grown vain in his conceit. He has discarded animal products and feels that he has found dietetic perfection in vegetables, grains, cereals, legumes and tubers. Scientific investigation and his own deficient physical and mental condition prove that he is no better off than his flesh-eating friends. Unbiased investigation shows that the modern Vegetarian is travelling the wrong frail. The things he eats are not natural products, nor the natural food of man. They are definitely artificial.

1. *Grains and cereals have been developed from insignificant grass seeds. By long ages of seed selection, careful breeding, intensive cultivation and constant fertilization, small grass seeds were developed by this artificial process into the modern grains and cereals.*
2. *Beans, peas, lentils, cabbage, lettuce, celery, etc., were developed in the same way from the grass family.*
3. *Tubers, which include potatoes, onions, carrots, turnips, beets, radishes, etc., are nothing more than wild weed roots, developed by the artificial process above described.*

Years ago Knight, in his "Vegetable Food of Man," asserted that grains and cereals have been developed from insignificant grass seeds now unknown to botany. Dr. Emmet Densmore, in his "Natural Food of Man," wrote: "Grains are the product of the temperate zone, not of those regions where there is no winter; and it was therefore a necessity of man's sustenance when he was without agriculture, without tools, and without fire, and had to depend upon foods spontaneously produced by Nature: that he live in a region where foods were produced at all seasons of the year. This narrows or confines the inquiry of natural food to two articles — fruits and nuts" (P. 224).

Vegetarians hate to hear their diet criticized, and most of them refuse to read or believe anything that is not favorable to vegetarianism.

C. C. Hibbs, D.D.S., wrote an excellent article on Dental Decay, in which he said: "Grains are responsible for nearly all of man's disease, for wheat, barley, oats and rye are no more a part of his food than oranges the food of the cow, or grains the food of dogs and cats. Man's food consists of the fruits and nuts of the tree. They are beautifully wrapped and hung on trees where the common herd cannot get to them. Man is given hands with which to remove this food and its wrappings, and eat to his content and perfect health. Eliminate grains from man's food and the decay in children's teeth will cease. Tartar and pyorrhea will disappear. The hospitals will fold up, and medicine will be a dream. All the doctors on earth and their 'vast' medical experience cannot disprove this statement. The medical profession dare not take a group of children and feed them according to Nature's law for a period of six months, and then truly publish the results" (You Can't Eat That).

Cereals Are Bad Foods

A German physician of note, Dr. Winckler, after enthusiastically adopting a vegetarian diet, was horrified to find in time that his blood vessels showed signs of cretaceous degeneration. With natural solicitude he applied himself to a solution of the phenomenon. He said he found the explanation in a work by Dr. Monin, of Paris, who, in turn, had been directed to the explanation of atheroma (arteriosclerosis, hardening of the arteries) by Professor Gubler, of Paris. The substance of the explanation was, that the mineral salts in the vegetables and the salt and seasonings used on the vegetables to make them palatable, are bad not only for their early effect on the body, but also because they induce the desire for, and the practice of taking, other and stronger stimulants.

In our work, "SCIENTIFIC LIVING," we mentioned the case of Captain Diamond who, as a vegetarian for 30 years, proved on himself that vegetables are not the panacea they are believed by many to be. His diet as a vegetarian consisted of grains, cereal products, tubers and green vegetables. At 79 he was a chronic invalid, suffering from a serious state of hardened tissues and blood-vessels, with stiffness of the joints. The muscles of his legs and back were so stiff that he could not sit down nor rise from a chair without great discomfort, and he often needed the aid of an assistant. The tissues of his hands and arms were so stiff that it was hard for him to hold his knife and fork to feed himself. Orthodox doctors were unable to help him and pronounced him incurable. They told him that he could not live long. He turned to Nature, became a fruitarian, and recovered health sufficiently to outlive all the doctors who gave him up, dying at the age of 120.

Diamond's and other cases of vegetarians supply data showing that vegetarianism is not what it is supposed to be. Grains and tubers contain large quantities of mineral salts that harden and stiffen the tissues, blood-vessels and joints, "and as a class," writes Densmore, "are the worst adapted as food for man." Dr. Rowbotham adduced proof in his work published in 1841, showing that "cereal foods tend to ossification (hardening) of tissues and joints, and produce decrepitude and early death" (Natural Food of Man, P. 390).

Fruits Easier Produced With Less Labor

Dr. G. Monin, of Paris, wrote: *"A vegetable diet ruins the blood-vessels and makes one prematurely old, if it be true that man is as old as his arteries"* (P. 312).

All authorities, both sacred and profane, agree that man was a frugivorian before he became a vegetarian. They also show that not only are fruits easier produced with less labor, but that any given portion of land is capable of producing more human food, with less labor, in the form of fruit, than under any other mode of culture. Down through the ages the body has been forced to adjust itself to everything man eats. Every herb of the field contains substances that are poisonous to him who has never eaten them. Not poisonous enough generally to kill him instantly, but often poisonous enough to make him sick even now, after having eaten them for thousands of years. Certain foods still give some people diarrhea, yet they have eaten them all their days — to do that they must be poisonous to the body. Green corn, roasting ears, often give people loose bowels. Strawberries, tropical mangoes, and other substances often cause a rash on the skin.

The fluid from the stem-end of a mango will poison some people as poison ivy will. If the wind blows through the damp foliage of a mango tree into the face of some folks for a sufficient length of time, their face will swell until their eyes are almost closed. A child of three years recently suffered from skin rash on much of its body from handling mangoes and not eating them. The mango is considered one of the finest of tropical fruits. When foods cause skin rash, some doctors order the eating of them stopped, while others say the rash is the result of a purging process that the body needs. The rash is similar to that of ivy poisoning, so why not be consistent and hold that the rash of ivy poisoning is the result of a purging process?

Vegetarians who read this will wonder what to eat. Melons contain no poisonous element and much distilled fluid that is good for the body. Tomatoes are not so good, but are better than many other vegetables because of their large liquid content. Man should return to berries and fruits and become a frugivorian, and then to fruit juices as he progresses toward Breatharianism by constantly reducing food consumption and dissipating the hunger sensation.

The press recently reported the case of a man who has been unconscious for four years as the result of a car accident. During this time the nurse gave him a liquid diet. For that time he was a Liquidarian.

Could man perform what is termed manual labor on such diet? What we call manual labor is unnatural. Wild animals do not engage in it. God did not intend man should chop wood all day, shovel coal or dig potatoes or post holes, and come in at night so tired he can hardly drag one foot after the other. Such labor is not natural and is degenerating. Spencer said that Perfect Correspondence must prevail. One thing calls for another to balance it. We must apply the Law of Correspondence to everything. In the word "Adaptability," lies the secret of degeneration and regeneration, advancement and retrogression.

The Law of Vital Adjustment makes man correspond always with his work, his habits, and his environment. Adaptability is the quality that makes this possible. Man's existence depends on the flexibility, pliability and adaptability of his body in order that it may be able to accommodate itself to new and varying conditions. The state of correspondence brings into operation the process of Vital Adjustment, and the purpose is to make man's body harmonize with his habits, his work, and his environment. These things become simple when the basic principle is understood.

Alimentation And Decrepitude

Parallax in his "Patriarchal Longevity," Easton and Bailey in their "Records of Longevity," Hufeland in his "Art of Prolonging Life," and Evans in his "How to Prolong Life," all substantially agree that the CAUSE of decrepitude, called Old Age, is not the work of Time, but the result of ossification and the deposition of calcareous earthy matter in the body — and it comes from what man eats, drinks and breathes.

What are the causes of the difference between youth and old age? Why do the functions of the body weaken and the organs deteriorate? Why does man become decrepit and die?

In Old Age there is a fibrinous, gelatinous and earthy deposit in the body. The solid earthy matter which, by gradual accumulation in the body, brings on ossification, rigidity, decrepitude and death, is composed chiefly of phosphate of lime, carbonate of lime (common chalk), and sulphate of lime (plaster of Paris), with magnesia and traces of other earthy substances. A process of solidification begins as soon as the infant begins to eat, and continues without interruption until the body is changed from a comparatively fluid, elastic, and energetic state, to a solid, earthy, rigid, weakened condition, which terminates in death.

Infancy, childhood, youth, manhood and decrepitude are so many different conditions of the body, or stages in the process of solidification or ossification. The only difference in the body between youth and Old Age, is the greater density and rigidity, and the greater proportion of calcareous earthy matter that enters into its composition.

Earthy Salts Cause Old Age

Common table salt, so freely used in the preparation of almost every kind of food, contains a very large amount of calcareous earthy matter; and produces great damage to the body economy. The theory is that the gradual accumulation of earthy salts in the system is the result of Old Age. Investigation shows that it is just the reverse. It is the cause of Old Age. If the number of years man lives causes the ossification that accompanies Old Age, then, as like causes produce like effects, all of the same age should show the same state of ossification. Investigation shows this is not so. It is common to find people of 50 who are as weak and decrepit as others are at 75 and 80.

Years ago an English doctor wrote: "Age is an evil that is not at all inevitable. It is not a question of dates and birthdays, but a matter of natural tendencies and dispositions. The age of the body is irrelevant so long as its condition remains young. Youth is not a time of life, but a quality, a trait of character, a mental and physical state."

In youth, the organs and structures are elastic, pliable, and yielding; the senses are keen, the mind active. In Old Age these qualities are usurped by rigidity, ossification. The senses are lacking in susceptibility, the mind in memory. In Old Age the arteries have thickened the walls and are smaller in caliber, caused by fibrinous, gelatinous and earthy deposits. Thus, the blood supply to the organs grows less and less; hence, their deterioration and the failing of their functions. Hardening and thickening of the blood vessels are not the work of Time, but of the earthy matter deposited by the blood — and the condition increases year by year, unless one changes one's habits.

Analysis shows that the blood holds in solution the earthy salts, the calcareous and osseous substances of the same kind as the solidifying agents, and arterial blood contains more than venous. This shows that each cycle of the blood leaves deposits of these damaging agents. It is the common carrier that clogs the system. But its supply must be replenished. Whence comes the new supply? From air, food and drink, from drugs and medicines. There is no other source.

Fruits Have Little Earthy Matter

We are dealing here with vegetables and cereals. Water and air will be noticed in due course. Writers have little to say about water in this respect, and nothing about air. Evans writes: "If man subsists on food that contains a large proportion of lime, a large proportion will enter into the composition of the chyme, the chyle, and the blood; and as from the blood the deposition of lime takes place, the greater the amount of lime that the blood contains, the greater will be the amount deposited in the system, the greater the degree of ossification, and the sooner will be produced that rigidity, inactivity and decrepitude which make him old and bring him to premature death. On the other hand, if the food and drink are selected from the articles that contain the least amount of lime, the least amount will enter into the composition of the chyme, the chyle, and the blood, the less amount there will be to deposit, the less degree of ossification, the less the rigidity, inactivity, and decrepitude, and the longer the life of man."

Dr. Evans says: "The cereals constitute the basis of modern man's food. They contain large quantities of mineral matter and, as a class, are the worst adapted as food for man, in regard to long life. Bread, man's so-called 'staff of life' is, to a great extent, the cause of his premature death" (P. 290).

Evans gives over twenty pages of tables of the analysis of foods, which show that fruits and nuts have the least proportion of earthy matter, as compared with their nourishing properties, of any of the foods now used by man. Next in order are animal foods, then come vegetables, and last are the pulses and cereals, which are shown to have the largest amount of earthy matter. Evans then observes: "From the foregoing analysis we see that fruits, as distinct from vegetables, have the least amount of earthy matter. Most of them contain a large amount of water, but that water in itself is the purest kind — a distilled water of Nature."

After quoting many authors on the subject, Dr. Evans says: "We have traced to the blood these earthy compounds that are found in the system, and which increase as age advances. By the process of transpiration, they are gradually, deposited by the blood. From the blood we trace them to the chyle, from the chyle to the chyme, from the chyme to the contents of the stomach, and thence to the articles of diet. Thus we eat to live, and we eat to die."- (How to Prolong Life, P. 28).

Evans filled many pages to show that food hardens and clogs the tissues and blood vessels, causing decrepitude and death. That is excellent evidence to prove that eating is not natural. If we grow decrepit and die because of something we do, then we should not do that thing. In the light of late nutritional discoveries, it appears more correct to say —
As we eat to live, we actually eat to die.

Fresh Fruit

So long as man must eat, the best food is fresh fruits, berries and melons. These contain fluid of the best and purest kind, distilled by natural processes. Some of these fruit juices are a wonderful solvent, opening the way into capillaries already clogged and hardened, provided the process has not gone too far. One author says that with a course of fresh grape juice, people with sunken eyes, wrinkled skin and poor complexion have made surprising improvement in recovering a younger appearance.

Lesson No. 17
Carnivorism Is Bad

"In nature a curious yet simple phenomenon is often observed — a rise and fall. If perpetual, it alternates and becomes a fall and rise. Man has degenerated. This degeneration is due solely to his diet. He has fallen; but we hope that he has risen to the highest point in the art of shortening his days, and that in the present generation he will commence gradually to fall back on his original and ordained diet. Since the creation, the days of man's existence have been little by little decreasing — it has been a gradual fall; but both science and religion tell us that he must rise again, that his life on earth must be prolonged." — Dr. Charles W. De Lacy Evans in "How to Prolong Life," 1879.

Evans made that statement 72 years ago. Were he alive today he would see that man has not yet commenced to fall back on his original and ordained diet — fruits and nuts. He is also in error when he says that man's degeneration is due solely to diet. For more than half a century we have read book after book on food and feeding, and have closely followed the explanations and arguments. We found that those who favored Vegetarianism omitted all the bad features, and the same course was pursued by those who favored Carnivorism. Books favoring Vegetarianism are composed by prejudiced authors who say nothing of the damaging qualities of vegetables and cereals. Those favoring Carnivorism are composed by authors who carefully omit the damaging properties of flesh. Such authors do much damage by giving their readers half truths. A half truth can be more dangerous than a lie, as it is more misleading. We will be soundly criticized by all hands for giving the facts as we find them.

One author writes: "The food of the natives of New Zealand and many South Sea Islands consists of fish, flesh, fowls, eggs, fruits, berries, leaves and sea-weeds, all of which contain a comparatively small amount of earthy matter. They are healthy and energetic beyond the age of 100 years, and are said to be equal to the finest young men in Europe after they have reached 100 years of age" (Densmore, P. 268).

We do not question these statements. But it is important to note that the instances of longevity mentioned occurred in the cases of those who lived in regions remote from the centers of civilization, and are free of the degrading influences and polluted air of such centers. We shall observe in due course some of the unnoticed dangers of polluted air that are striking people down in civilization at an alarming rate. Those South Sea Islanders live a more natural life, breathe better air and subsist on a diet that damages the body less than the conventional diet of civilization. If these natives, in their favorable environment, ate only fruits and berries, it would no doubt double and perhaps triple their present lifespan.

Butter, Milk And Cheese Less Harmful

The early Greek historian Herodotus told of a people of Ethiopia who, because of their unusual longevity, were called Macrobians. Their diet consisted entirely of roasted flesh and milk, both of which contain only a small amount of earthy matter. They were remarkable for their "beauty and large proportion of their body, in each of which they surpassed other men," he wrote. They lived to be 120 and some to be much older (Densmore, P. 268).

Fishermen and those living near the sea who subsist chiefly on fish, have good health and live to considerable ages. Fowls that subsist chiefly on fish and flesh, as the pelican, vulture, hawk, eagle, owl, have much longer life-span than domesticated fowls, as chickens, turkeys and pigeons, fed large quantities of grain.

According to Captain Riley, some tribes of Arabs of the desert, subsisting entirely on the milk of their camels, have no sickness nor disorders, and attain to great age, with remarkable vigor and vitality. He wrote: "I am fully of the opinion that a great many Arabs on this vast desert live to an age of 200 years and more. Their lives are regular from birth to death; their climate dry and unchangeable; they are not subject to hard labor, yet have sufficient exercise for the purposes of health."

Camel's milk contains little earthy matter and does little damage to the body; the people are not subject to the hard manual labor of the toilers of civilization; the climate of their region is not changeable; their lives are regular, and, more important, the Breath of Life they inhale is not polluted with the poisons of civilization. John Smith cites the case of Ephraim Pratt, of Shutesbury, who died in 1804 at the age of 116. For forty years he lived very much on milk, and yet he could "mow a good swath" almost up to the day of his death (P. 275). Smith mentions the case of "Paul the hermit" who lived to be 115. He spent nearly a century in the desert, and lived largely on dates and water. Also the case of a shepherd at Gompus, Hungary, who lived to be 126. He "subsisted entirely on milk, butter and cheese, and was never ill" (P. 277). But in those days the devitalized fluid termed "pasteurized milk" was unknown.

While the articles of diet such as milk, butter and cheese are less harmful than many other things men eat, they are not suitable for the body. Science finds that one cause of hardened blood vessels and tissues is the cholesterin contained in milk, eggs, butter and cheese.

One scientist says: "Hardening of the arteries, which has been experimentally produced by other agents such as a high protein diet of whole wheat, is undoubtedly due chiefly to the deposit of cholesterin along their interior walls. "Also, it is an interesting fact that just as the excess of cholesterin in the senile organism causes the characteristic symptoms of arteriosclerosis, so it causes the increased incidence of cancers, tumors and carcinomas at the stage of life. Every other animal except man is weaned in infancy, but man continues drinking milk or using it in foods all his life. Very few animals eat eggs, and then only during the three or four weeks egg-laying season, but man eats them all year round all his life. Cholesterin is an animal fat found in meats, fish, fowl, eggs, lard, butter, milk, cream and cheese. Hence, the first offensive in the attack on all disorders of old age must be a diet that excludes the animal products just mentioned."

The paradox that man eats to live and eats to die explains itself as we proceed. A child can understand that if a man live 200 years on a diet of camel's milk, or dates and water, and only 50 on a diet of flesh, cereals and vegetables, the difference in the life-span depends chiefly on the difference in the diet and the other factors enumerated, such as climate, air, labor. We should not make the mistake of giving diet all the credit, as most writers do. Under the same circumstances, it is possible for man to live 200 years along with these Arabs, eating the foods of civilization, provided he keep down to a minimum the amount consumed so the body is able to handle it. In due course we shall cite cases of men who have lived up to 256 years, and their great ages did not depend entirely on the food they ate.

Reason For Increased Vitality

Now for some facts concealed by the authors who favor Carnivorism and believe that a diet of fish, flesh and milk is responsible for the remarkable longevity of those who subsist on these substances. Animal food possesses a greater proportion of stimulating power to its quantity of nutriment matter than vegetables and cereals. For that reason it accelerates all the functions of the system, rendering vital changes quicker and less complete, and the general result of the vital economy less perfect.

Moore demonstrated at the Harvard Laboratories of Physiology that a diet of flesh produces acceleration of heart action that is surprising in its magnitude and duration. After a meal of meat, the increase in heart rate regularly amounts to 25 to 50 per cent rise above the fasting level, and persists, in experimental subjects, for 15 to 20 hours, to reach a total of many thousands of extra heart beats. As the heart beats in harmony with the blood flow, this shows how much faster a stimulating flesh diet makes the blood move and the body organs work, as their work keeps pace with the blood stream. The findings of Moore were confirmed by Dr. Arthur Hunter, actuary of the New York Life Insurance Company. His investigation showed that flesh eating quickens body function, heart action, and raises blood pressure. It requires the presence of internal poisons to cause the body functions to quicken in this manner. The stimulating effect that appears to result from flesh, rises from the rapid decay of the flesh. The flesh decaying in the digestive tract forms some of the most deadly poisons known to chemists. The body intelligence senses the danger and knows the poisons must be eliminated quickly to minimize the damage to the tissues and organs. So the speed of all functions is increased to cast out the dangerous enemy as soon as possible. We now meet another paradox. The quickened functions make the man feel stronger, and he gives credit for it to the flesh he ate. He is right, but how ignorant he is of the price he pays for that temporary increase of vitality.

There is no law of organic life, extending over the whole animal and vegetal kingdom, more general and more certain than this: The slower the growth of organic bodies, consistently with the healthy and vigorous condition and action of the vital powers, the more perfect and symmetrical is the general development. In the vital economy of the human body, all the changes concerned in the development and maintenance of the system are the most healthfully slow and complete when the food consists of fruits. It follows from every known physiological principle in the human constitution, that — all other things being equal — a diet of fruit is most conducive to the completeness of bodily development and perfectness of symmetry and beauty.

Flesh Foods Putrefy

The rapid transformation of the tissues in the carnivora is a condition of their existence. It is only as the result of the change of matter in the body that its vital organs are the better protected from the damage resulting from the poisons generated by decaying flesh in the bowels and blood. Another fact is worthy of observation in connection with our subject. So far as chemical tests are concerned, the chyle of all animals is the same, from whatever kind of food it may be formed. But as to its physiological qualities and its relations to the vital economy, its character varies with the foods consumed. Physiologists unite in stating that chyle formed from flesh food will putrefy in three or four days at the longest, while chyle formed vegetable food, because of its greater purity, may be kept for many days without becoming putrid. They find that human blood formed of flesh food will putrefy, when taken from the living blood-vessel, in a much shorter time than that formed of vegetal aliment; and that, other things being equal, there is always a greater febrile and putrescent tendency in the living body of these who subsist largely on flesh, than in those who subsist wholly on vegetables.

If two healthy, robust men of the same age, the one subsisting mostly on flesh and the other exclusively on vegetables and water, be suddenly shot and killed in warm weather, and both bodies be laid out in the usual manner, and left to the action of the elements, the body of the vegetarian will remain two or three times as long as the body of the flesh-eater, without becoming intolerably offensive from the process of putrefaction. This fact was fully confirmed by Majendie. It may here be noticed that the excretions from the lungs, skin, kidneys and alimentary canal of the Herbivora are far less offensive than those of the Carnivora. We know from this that the breath, perspiration, body odor, etc., of the vegetarian are not so unpleasant as those of him who eats flesh food.

From the foregoing facts, it may be concluded that the more rapidly the changes in the chyle of the Carnivora, the more rapid is the state of ossification, hardening, and stiffening of the body. All processes of bodily decay are accelerated and the approach of Old Age is hastened. Hence, a diet of flesh is less favorable to health and longevity than one of fruits and nuts, which form chyle and bloodless subject to chemical decomposition, and require less rapid changes.

Hufeland said: "The more slowly man grows, the later he attains maturity; and the longer all his powers are in expanding, the longer will be the duration of his life. It is natural law that the life of a creature is lengthened in proportion to the time required for growth and development."

Mode Of Living Builds Cravings, Aches And Pains

Most of the stimulating effect of flesh food rises from its rapid decay in the stomach, bowels and blood. The body strives to protect itself by an increase of its functions in order to eliminate the poisonous products as quickly as possible. As this is continued year after year, the victim is hurried to the grave. Haig said, "Seventy-five per cent of the most terrible disorders from which men suffer, rise from poisonings of unnatural food. In a way that there is no misunderstanding, Nature says that man is a frugivorian and not a carnivorian."

Another authority states that flesh as food is more dangerous for man than nicotine, because nicotine is a single poison while flesh contains eight dangerous poisons, He adds: "According to the universal statistics on mortality made by Westgaards, in England, where flesh is the people's staple article of food, only one in every 100,000 lives to the age of 107. Of 100,000 new-born children, 30,000 die in their first year and 11,000 in their second. This extraordinary high rate of infant mortality has only one cause: The complete intoxication of the mother's organism by the poisons in the flesh food, and by her inferior liquids which poison the child while still in embryo." — Cosmotherapy, P. 284.

The flesh-eating man is in a constant state of mild irritation, intoxication, rising from the excessive stimulation caused by the poisons of the decaying flesh food in his stomach, bowels and blood. This man has been a flesh-eater from childhood, and the condition is deep-seated and chronic. His body and organs are adjusted to it. As his nerves and brain begin to be comparatively free of the intoxicating effects, of course he does not feel right. The brain and nerves begin to call for their regular stimulant; they are used to it, adapted to it, and crave it. If the flesh and coffee and toast come not at the accustomed time, they protest.

Alcoholic intoxication is that state in which the poison dulls the brain and nerves. Polluted air produces the same deadly effect. So does any poison if it is powerful enough, or enough of it is taken. The flesh-eater, coffee and coke drinker, the smoker — they are all in a state of chronic autointoxication. When deprived of their indulgences, the effects begin to wear off, and the nerves wake up. Then trouble starts. If the condition of autointoxication is mild, the symptoms are mild. If advanced, the symptoms are stronger, and give the victim more trouble. The sensations of hunger, uneasiness, nervousness, weakness, are the symptoms of the protesting nerves as they begin to revive from their chronically poisoned state. The trouble frightens the victim and he sees his doctor, who administers another poison to dull and weaken the protesting nerves. That stops the symptoms and "cures" the patient.

When the body becomes adjusted to this chronic condition, it cries out for the stimulating poisons which keep it in that state. It craves them; it must have them. The nerves protest when the stimulating poisons are not forthcoming. Then a substitute is given in the shape of drug and serum poisons. We build the world in which we live. To change our world we must change ourselves. Our mode of living builds our cravings, aches and pains. We should know our bad habits and conquer them. We should know what to eat and drink, how to live in harmony with cosmic law — and live accordingly. There is no substitute.

48 Million Have Trichinosis

We have heard of trichinosis. Trichinella Spiralis is a parasite that imbeds itself in the muscular tissues of animals in the form of encysted worms. They infest dogs, cats, rats, hogs and other animals, live on garbage and decaying animal and vegetal matter. The worms get into the bodies of people who eat pork not cooked sufficiently to kill them.

A recent article in the Therapeutic Digest stated that, through a special technique that shows up trichinosis, post-mortems have shown that in Cleveland 35% of the inhabitants had trichinosis when they died; 24% in Washington; 31% in Minneapolis and Rochester; 43% in San

Francisco; 49% in Boston; and that 48 million people in the nation have trichinosis. If you must eat flesh and do not want worms working in your body, have your flesh well cooked to kill the worms. It is better to have the worms dead in your body than alive. We do not recommend flesh eating. But as one moves back to a diet of fresh fruit, one may use unpasteurized milk, cheese and butter, without salt, until these may be discarded entirely. These substances are products of the vital processes, and do the body less damage than vegetables and cereals. No salt should ever be used on anything.

Henry S. Graves, U. S. Forest Service, wrote: "To primitive man the forest furnished both food and shelter. Later, when he became a meat-eater, he left the forest for the treeless plains, where he found in abundance the animals upon which he preyed" (Mentor, June, 1918).

One writer says, "During the Moon Period man was fed upon the milk of Nature. Cosmic food was absorbed by him, and the use of the milk of animals has a tendency to put him in contact with the cosmic forces." The real Milk of Nature is that contained in fresh fruits, berries and coconuts. Man should change to this diet as soon as practicable, and then move on to Breatharianism by gradually reducing the amount of food and liquid ingested, and getting into the good air of the hills and forests.

Table Salt

Common table salt is a compound of sodium and chloride, a mineral in inorganic form that cannot be used by the animal body. It enters the body as salt and leaves it as salt. Salt is a deadly irritant to all the tissues of the body. Put some salt in your eye and feel the distressing effect. Salt in food irritates the membrane of the stomach, and for protection, mucus is excreted by the cells. The salt passes to the bowels, the membrane of which pours out more mucus for protection, in time a catarrhal condition results. Any substance that irritates the body cells, causes the mucus membrane to excrete mucus for protection, and creates a catarrhal condition in time, no matter whether it be salt, polluted air, pepper, vinegar, spices or any other irritating substance. Salt irritates the cells and they call for water to allay the irritation. This creates abnormal thirst and results in a water-logged body.

As the salt irritates the cells year after year, the cells and tissues harden, the blood vessels harden, the blood pressure rises, with its train of troubles. Some of the salt is filtered from the blood by the kidneys, and in time the irritation of the salt creates a condition in the kidneys termed Bright's disease. Haig proved that salt impedes the elimination of uric acid, which thus paves the way for gout, sciatica, rheumatism, lumbago — all symptoms arising from the use of salt and treated as "diseases." As you get older the flavor of food changes, because the use of salt, spices and condiments has dulled the delicate taste buds of the tongue and weakened your sense of taste. Salt eaters say unsalted food tastes "flat." It depends on what one is use to. No carnivorous animal in its native state uses salt, except as an acquired habit. Animals form bad habits as man does.

Opinions On Salt Eating

The North American Indians used no salt when discovered by the Europeans. Chinese in the interior of their country use no salt. Most of the human race that subsists chiefly on vegetable food uses no salt.

Dr. J. E. Cummins wrote: "I knew of a case of a little girl who had a craving for salt. She would take a teaspoonful of it at a time when not watched. She was a pinched-faced little thing, and had hardening of the arteries, was wrinkled and appeared old at the age of four years."

Commenting on how salt dehydrates animal flesh, Professor Liebig said: "Fresh flesh, over which salt is strewn, is found swimming in brine after 24 hours, yet not a drop of water has been added. The water has been yielded by the flesh itself." Dr. Bouchon observed: "Salt is one of the worst of social poisons. Because of its use, surgeons are constantly operating for appendicitis, gastric ulcers, and liver and kidney calculus. It atrophies, dries up or hardens the tissues, and causes persons with tendencies to arthritism to become stout, and those of lymphatic temperaments to become thin." — Nouville Review.

Dr. Hal Bieler stated: "Haig showed that in animals, such as dogs, and in fowls, such as chickens, where a good deal of nitrogen is eliminated as uric acid as the result of feeding salt, even in very small quantities, the creatures soon die. Autopsy showed the liver and kidneys studded with uric acid concretions. Our forefathers used a salt solution as an embalming fluid. The ancient Egyptians used oils, spices and salt in their mummy wrappings. Today, we mummify the living with salad dressings made of mineral oils, spices and salt. You see these mummies walking the streets. The dry skin, shrunken bodies and faded hair bespeak the hardened livers and sclerotic kidneys. It is hardly necessary to embalm such bodies after they are dead, for they are already pickled to the gills. "The action of salt on hogs, rabbits, etc., is to paralyze the muscles of the hind quarters and the animal sinks to the ground. Later, the muscles of the lungs are paralyzed and the victim dies of asphyxia. If this is the result of the consumption of salt by animals, it is reasonable to suppose that salt has a similar affect on Man." — Philosophy of Health.

Dr. A. Birchard said: "Man is the only animal that deliberately commits suicide by self poisoning. He is the only animal that spoils his food before he eats it. The average individual suffers constantly from chronic poisoning of some kind, due to the food he eats, either in wrong combination or in excessive amounts, or by adding to its injurious substances to stimulate a jaded appetite. Instead of fasting until he has no appetite for the simplest of foods, he tries to whip up an appetite by the use of irritating and injurious condiments, or he doses himself with poisons of various sorts. He begins the day perhaps with a poison dose, in the form of coffee, to wake him up, or maybe a drink of whisky or bitters to get an appetite. Probably he finds an afternoon cup of tea necessary to relieve after-dinner stupor. At night he needs a narcotic to put him to sleep, and in the morning a cathartic to move his bowels. With all his other poisonings, he spoils his food by putting into it toxic substances which, by means of acrid, biting and burning flavors, belong to the poison class and are not fit to eat. These poisonous substances, used for their flavoring properties and having no food value, are known as condiments." — Philosophy of Health. It seems strange that the doctors who know these things, will say that it is more difficult to explain why man dies than it is to show that he should live forever. One eminent author states: "Wild boars and other mammals exceed the age of 200 years because they instinctively follow a natural diet. It is only man who eats everything indiscriminately. He arranges his meals by the clock (due to the artificial life he lives), and so his hunger is only a matter of habit and not natural hunger" — Cosmotherapy, P. 280.

Ancient Wisdom

The fourth or highest plane in which man can function is called the mental world. This mental world is the mind body of the solar God. The (Ancient) Mystery Schools by means of four initiations, teach man how to Junction consciously in the four worlds of Nature. In the fourth initiation, they teach him how to use the little area of consciousness, which he calls his mind, as a vehicle by which he can function consciously within the mental body of the Grand Man. In other words, he is taught how to wander around in the mind of God. This may seem a very peculiar idea, and yet, the system of accomplishing this has been taught by the Egyptians, Chinese, Hindus. Chaldeans, and early Christians for thousands of years. — *Manly Hall, Super Faculties and their Culture*, p. 28. "When any object or purpose is clearly held in thought, its precipitation, in tangible and visible form, is merely a question of time. The vision always precedes and itself determines the realization." Lillian Whiting.

"Compared with what we ought to be, we are only half awake. We are making use of only a small part of our mental and physical resources. Stating the thing broadly, the human individual thus lives usually far within his limits; he possesses powers of various sorts which he habitually fails to use."

— William James —

Lesson No. 18
Longevity

"There is a vast difference between the longevity of man and that of animals. If the length of a stag's life were one year, a man should live for thousands of years. All these animals live for centuries, so, according to cosmic law, man should live for some thousands of years.... If we fast two days a week, then eat only fruit and obey the other laws of life, we can approximate to the longevity of the biblical Patriarchs." — Professor Edmond Szekely in his Comsotherapy.

In the preceding lesson we stated that it is possible for man to live 200 years along with those certain Arabs mentioned, while eating the damaging diet of civilization, provided the amount of food consumed be kept down to a minimum so the body has time to handle it; and provided further that man live in the same favorable environment and in the same manner in all other respects as these Arabs.

In our years of writing on these subjects, we have collected a large number of cases from many sources of people who have lived from a century to 370 years. Many of these not only consumed the damaging diet of civilization, but some of them breathed the same polluted air of civilization. That makes their advanced ages the more amazing when we learn how seriously polluted air damages the body. Mehlis cites the case of a woman of 96 who was unable to eat for eight months except a little water because of the persistence of her illness. Her teeth grew again, her hair became darker and thicker, and she looked young again. She lived in good health for another 23 years. We saw in Lesson 11 that the Hindu lady of 68, who had eaten nothing since she was 12, was "always gay and looks like a child despite her age."

In the press of 1931 Robert Ripley stated in his "Believe It or Not" that J. D. Cameron, of Augusta, Maine, could shoulder a barrel of potatoes when he was 100 years old. In his "Believe It or Not' published in 1933, Ripley stated that Harriet Breedlove of Tennessee cut a new set of teeth at the age of 102; that Thomas Gordon of Michigan had his hair turn its natural color at the age of 103; and that Daphne Travis of Georgia cut a third set of teeth at the age of 108. The press of June 7th 1949, reported the death of John H. Gates at the age of 104. He was one of the three remaining Union Army veterans in Ohio. The press of May

18th 1949, reported the death of Robert M. Rownd, age 104. He was Junior Vice Commander of the GAR of New York State.

The press of May 27th, 1949, carried a picture of Joseph Manning, who was celebrating his 104th birthday by dancing with a young woman. Margaret Krasiowna, of Poland, died in 1768 at the age of 108. She married her third husband at the age of 94 and bore him two boys and a girl. The third husband died at the age of 119. The press of February 6th 1919, reported that in September 1875, a couple named Ballat climbed to the top of the Column Vendome, the husband being 110 and the wife 106.

The press of September 6th 1947, stated that Jesus Andasole, of San Jose, California, 110 years old, believed he was starting life over again. His hair, grey for years, was turning black again, and he was cutting his third set of teeth. James A. Hard was born in Victor, New York, in 1841, enlisted in the 37th New York Voluntary Infantry at the age of 19, fought in the battles of Bull Run, Yorktown, West Point, Fredericksburg, Chancellorsville, South Mountain and Antietam, was discharged from the army June 29, 1863, and the press of August 26th, 1951, reported that "when he passed his 110th birthday a few weeks ago, he was smoking his cigar, eating all his meals, feeling quite chipper and ready to comment on local, national and world affairs." The press of August 19th, 1951, reported that Leonard Finch, of Panama City, Florida, "A spry oldster of 111 years," didn't get his first airplane ride today "because his son objected, believing the excitement might be too much for his father, who declared that he expects to live a long time yet." The press of July 10th, 1951, carried a picture of Henry L. Hall, negro, who had just passed the age of 112, He admitted that he was a bit slower on his feet than when he was 100. The press of March 13th, 1937, reported the death of John Weeks, of New London, Connecticut, at the age of 114. When he was 106 his hair turned its natural color, new teeth appeared, and he married a girl of 16. His diet consisted mostly of baked beans and corn bread, had he remained single or married a woman of 70 or 80: he might have lived longer. The press of June 20th, 1938, reported the case of Sally Dollar, Cherokee Indian who had lived for 116 years on top of Lookout Mountain without ever coming down. Lizzie Deevers of Sapulpa, Oklahoma, was reported in the press of June 10th 1945, as being 114 years old, had been married nine times, and was on "a manhunt for Mr. No. 10."

Prof. Hotema   *Man's Higher Consciousness*   Lesson No.18

The press of January 26th, 1944, stated that Kate Williams, of Ocala, Florida died at the age of 117, and added, "Until about seven years ago she often walked from her country home to town, a distance of 19 miles." Hufeland mentions an old man who lost all his teeth when he was 117, and then grew a new set. He mentions the case of a man who fasted for several months at the age of 60, grew his teeth again, recovered his youthfulness, and lived for another twenty years.

The press of April 1st, 1945, reported the death of Mrs. Ramirez Trujillo of Riverside, California, at the age of 118. The press of February 19th 1945, reported that "Indian Ned Rasper" was 119 years old. He was born in Siskiyou County, California in 1826. The press of March 9th, 1927, reported that Charles W. Ellis, of Ada, Oklahoma, had just celebrated his 119th birthday; his mind was alert and memory good. He said he owed his long life to frugal eating, drinking water only, and living in the open air. The press of July 14th 1922, reported that Elsie Guest, a negress of Muskogee, Oklahoma had just celebrated her 120th birthday. She was well and vigorous and remembered the battle of New Orleans in 1812. John White, of San Springs, Oklahoma, was born April 10th 1816, in Georgia, spent 88 years in slavery, and in good health; he celebrated his 121st birthday April 10th, 1936.

The press of March 5th, 1932, stated that Mrs. Bell Ryans, the day before celebrated her 121st birthday. She was born March 4th 1811. Census records show that James W. Wilson, of Vidalia, Georgia, was born May 15th, 1825; his death at the age of 120 was reported in the press of December 25th, 1945. In his book in 1915 titled "Long Life In California," Dr. Thrasher referred to the Case of Captain Diamond as follows: "Captain Diamond, who published a book sixteen years ago, entitled 'How To Live to be 100, lives at Crocker Old People's Home in this city (San Francisco). The author has personally known him for 28 years. He was then 96 and today he looks no older."
Captain Diamond died when he was 120. His father lived to be 104. Senora Leandra Chairez, of Santa Ana, California, was 121 on September 26th, 1934. She had records showing she was born in 1818. On August 27th, 1931, Mrs. David Valvero, Of Sacramento, California, died at the age of 128. She married the last time when she was 120.

The press of July 20th, 1946, stated that Jasper C. Darrett, a negro living near Houston, Texas, died the day before at the age of 121. Antonne and Jacques Desbordes, brothers, were book dealers in Holland, and publishers of Voltaire's works. One died at 124 and the other at 125.

Prof. Hotema   *Man's Higher Consciousness*   *Lesson No.18*

The press of June 13th, 1922, reported that Peter Nedall, of Bulgaria, had just celebrated his 124th birthday. He still worked in the field and walked erect. Beans, porridge and sour milk constituted his diet. He never used tobacco, drank some milk occasionally, and never visited a dentist. The press of March 4th, 1928, reported the death of William Kennedy at the age of 126. He was born in Ireland and migrated to Canada when he was 105. The press of January 25th, 1923, reported the death of Eveline Booth, a negress of Atlanta, Georgia, at the age of 126. H. H. Glenn, register of births and deaths, reported the date of her birth as March 18th, 1797. Kiziah Hotato was an Indian girl of 15 and rode a pony over the historical "trail of Tears" in 1823, when the U. S. Government moved the Creek, Cherokee, Choctaw, Chickasaw and Seminole Indians from their developed homes in Georgia, Alabama and Mississippi to the wild region termed the Indian Nation, which became the State of Oklahoma in 1908. She died December 21st, 1934, at the age of 126.

Mrs. Martina De La Rosa, of Delhi, California, celebrated her 129th birthday November 12th, 1934. The press of April 13th, 1949, stated that "when the civil war ended Jose Garcia, of Victoria, Texas, was 45. He had taken an active part in Mexico's war with Texas in 1835. When World War I ended in 1918 he was 98, but still full of pep. Ten years ago he received wide publicity when he registered under the alien registration act, giving his age as 119. But he will witness no more wars, as he died yesterday at the age of 129." Pierre Defournel, of Marjac Vivirias, died in 1809 at the age of 129. He married his third wife when he was 120 and she was 19. In 1913 Anton Turitsch, of Heregovia, was living at the age of 131. He walked to church every Sunday, eight miles each way, and remembered the important events of the world's history for 125 years.

The press of August 20th, 1946, stated that James E. Monroe, of Jacksonville, Florida, a son of the 5th president of the U. S. A., was born July 4th, 1815, on the outskirts of Richmond, Virginia, and was 131 years old. He said that his advanced age was due to the fact that he had formed the habit of sleeping 15 hours at a stretch on the ocean beach. He was not afraid of that bad, damp, night air which the doctors urge people to avoid. The press of June 30th, 1922, reported the case of Jan Krasanski, a Pole, the only survivor of Napoleon's armies. He fought in the battle of Borodino at the age of 22, which made him 132 in June 1922. The report stated that he looked like a sturdy octogenarian. In 1943

Prof. Hotema   *Man's Higher Consciousness*   *Lesson No.18*

Sayed Mahrem of Chicago celebrated his 132nd birthday. He was born in Egypt in 1811.

The press of March 18th, 1943, stated that Santiago Surviate, an Indian, died March 16th, 1948 at the age of 134. The records showed he was born in Arizona in 1808. Ripley stated in his "Believe it or Not" in the press of February 6th, 1987, that Joseph Crele, born in 1726 near Detroit, was accidentally killed in 1866 at the age of 140. His hair turned black and he grew new teeth. Calcas, of Peru, died in 1761 at the age of 140. Hilario Pari, also of Peru, was 148 years old when seen by Humboldt. Up to his 130th year he walked 10 to 12 miles each day for exercise. In 1927 Domingo Jacinto, Chief of a tribe of Digger Indians in California, was living at the age of 144.

The press of February 25th, 1927, stated that in carrying out the 1927 census in Russia, census officials found nearly 150 persons who were more than 100 years old. The oldest was Ivan Shapkovsky, whose birth certificate showed he was born in 1728, making him 145. Among the oldest women was Martiana Maliarevtch, who had passed her 131st birthday. She walked 20 miles in the snow to the census once to be sure she was registered. Drakenberg, a Dane, buried in the cathedral at Aarhus, Denmark, lived 146 years, and was more often drunk than sober. At the age of 111 he married a woman of 60. Drunkards may reach a ripe old age, but gluttons never. Ripley stated in his "Believe it or Not" that John Haynes, a private under General Washington died at the age of 132; and that Kebenah Giveywence, a Minnesota Indian, died at the age of 151.

Ripley, in his "Believe it or Not," in the press of January 9th, 1932, reported the case of Martina Gomes, as "The most ancient woman of the Western Hemisphere," having then just died at the age of 153. Dr. Marion Thrasher, in his book "Long Life in California," stated that the Indians of southern California and Mexico, who subsisted on a simple diet of fruits, corn, acorns and vegetables, lived 120 to 150 years. He cited the case of "Old Gabriel," who died of pneumonia in 1890 at the age of 150.

Gabriel had lived on fruits, nuts and corn. He could thread a needle without glasses two years before his death. His hair held much of its natural color to the last. An autopsy showed his organs were in good shape. Had polluted air not killed him, he might have lived another century. Neils Paulsen, of Uppsala, Sweden, died in 1907 at the age of 160, leaving two sons, one 103 and the other 9 years old.

Prof. Hotema *Man's Higher Consciousness* *Lesson No.18*

Zora Agha was born in Turkey in 1774, and died in 1936 at the age of 162. He married 11 times and at the age of 96 became the father of his 36th child. He buried ten wives and 27 children. In the press of 1931 Ripley stated in his "Believe it or Not" that Christian Mentzelius grew a complete set of new teeth when he was 120. The case became famous in dental circles, and is known as the Menel case. It was attested to by Dr. Schengren, who was well acquainted with the circumstances. In 1922 Djouro Chemdine, of Turkey, was trying for work in a dime museum, his qualifications being that he was 164 years old.

The press of March 21st, 1942, stated that a "grey-haired colored man," docketed on a minor charge, calmly told the desk sergeant at Memphis that his age was 169. He said, "I was born in slavery and was over 90 when Lincoln was killed." He was born in 1773 and was three years older than the United States Declaration of Independence.

Henry Jenkins appeared in court as a witness in a matter that occurred 140 years before. Two sons were with him; one 100 and the other 102. He was born May 17th, 1500, in Yorkshire, and died in 1670 at the age of 170. He never ate cooked food and never ate in the morning. He lunched at noon on milk or butter and fruit. In the evening he had only milk and fruit. Janos Roven and his wife, Sarah, were married 147 years. They died in 1925, almost on the same day. He was 172 and she 164. They left a son 116. They subsisted on a frugal diet and ate practically no flesh. According to Voltaire and Francis Bacon, there appeared in Court in the reign of Henry IV and Louis X III a knight with every appearance, of physical and mental perfection, who looked like a man of 40. His name was the Count of St. Germain, and he remembered all the events of history covering a period of 150 years before. Of him Bacon wrote, "Whenever he was invited out to suppers and dinners, he touched nothing but fruits and only a very little of them and he sometimes fasted completely for several weeks." Joseph Surrington died near Bergen in 1797 at the age of 160. His eldest son was 103 and his youngest only 9.

The press of February 27th, 1988, reported the case of Yogi Tapsi Bishan Das Udasi, who was then 172 and appeared to be not over 40. Louise Truxo, a negress, died in 1780 at the age of 175. The Countess Desmond Catherine lived to the age of 145. She ate practically nothing but fruit. In 1878, Miguel Solis, half-blood Indian, of Bogota, San Salvador, was found by Dr. Louis Hernandez working in his garden. Solis said he was 180, but his neighbors said he was much older.

Hernandez was assured that when one of "the oldest inhabitants" was a child, Solis was recognized as a centenarian. He ate once a day, in the afternoon, and his food consisted of fruit and milk. He fasted the first and 15th of every month and was never ill. The press of July 24th, 1921, stated that Jose Calvario died at Tuxpan, Mexico, at the age of 185. Church records showed he was born in 1727, He was active up to the time of his death.

On his mission in Arabia, Dr. Weber noticed an old woman who ate but once a day and then consumed only a few dates. She was a strong woman and Weber thought she was about 40. He was extremely surprised to learn that she was 198, "despite her miserable diet." She told him that when she was 156 her teeth were renewed for the third time, and that all her symptoms of regeneration always appeared after a prolonged fast.

Kentigern founded the Cathedral in Glasgow and died at the age of 185. Pierre Zortay of Hungary died at the same age. Don Juan Saveris de Lima died in 1730 at the age of 198. A Russian soldier died in 1825 at the age of 202. According to the records of St. Leonhard's Church, London, Thomas Carn was born January 25th, 1588, and died in 1795 at the age of 207. Like Jenkins, he ate sparingly, and never ate cooked food.

His diet consisted of milk, butter and fruit. The appearance of Carn when he was 150 was that of a vigorous man of 50 or 60. He survived 12 kings Of England. Thomas Parr of England died in 1635 at the age of 152. He married at the age of 84, "seemingly no older than many men at 40." He was brought to London by Thomas, then Earl of Arundel, to see Charles I., "when he fed high, drank plentifully of wines, by which his body was overcharged, his lungs obstructed, and the habit of the whole body quite disordered; in consequence, there could not but be speedy dissolution. Had he not changed his diet, he might have lived another century." — Easton.

The celebrated Dr. Harvey, modern discoverer of the circulation of the blood, dissected the body and found every organ in perfect condition. Harvey said he found Parr's cartilages soft and flexible, and "his testes were sound and large." Harvey expressed the opinion that Parr could have lived for another century. Then it is possible for every normal man to live that long, as Cosmic Law has no favorites and treats all men alike. In his 102nd year Parr was found guilty of a misdemeanor, and facts were adduced at the trial which showed that this "man of 102 years really had the qualities of a powerful young man." (Lorand, in Old Age

Deferred). Shall we believe that this centenarian was found guilty of molesting some young woman? Sir William Temple wrote that the Brahmins of India, at the time that country was discovered by the Europeans, lived to a great age. Some who subsisted mostly on rice reaching the age of 200, while some in other parts of India, who ate chiefly fruit and green herbs and drank only water, lived to be 300 years old.

He Lived 256 Years

The St. Louis Post-Dispatch of June 11th, 1938, reported the death of Li Chung-Yuen, a Chinaman, at the amazing age of 256. The account was written by Keith Kerman "of the Post-Dispatch Sunday Magazine Staff," who said: "According to the popular account, Li was mature enough when the great earthquake of 1703 wiped out 200,000 Japanese, to refrain from undignified rejoicing, and he was about to become a centenarian when Washington crossed the Delaware. "A few years ago a professor in the Minkuo University reported that he had found records showing that Li was born in 1677, and had been congratulated by the Chinese government on his 150th and 200th birthdays."

Li stated in his lifetime that he was born in the 16th year of the reign of the Emperor Kang Hsi, and related many stories of his youth that appeared to prove that he actually remembered events that occurred during the regime of that long-dead Mongol Monarch, whose reign began in 1661, and who died in 1722. In further support of his claim of astonishing longevity, Li counted off 23 wives who had long since gone to the land from which no traveler returns. In 1827 the Chinese government sent an official felicitation to Li on the occasion of his 150th birthday; and in 1877 the government again by letter congratulated him on his 200th birthday. In May 1930, at the age of 252, Li was lecturing to the students at the University of Chang Fu. At the age of 209 he lectured twice each day, three hours at a time. Twenty-eight sessions in all were held. That task had taxed the energy of a man of 40, but Li left each lecture fresh in body and clear in mind.

Throughout the day Li behaved like a buoyant youth, who was enjoying the opportunity afford him to tell some 1500 of his listeners, whose ages ranged from 18 to 80, something of the secrets of longevity.

William M. Goodell says that he was in Canton in 1838 and heard considerable talk about Li, and learned that in the first century of his life he followed the occupation of a herb gatherer. He stated that Li "was a vegetarian who ate only herbs that grew above the ground, and fruits of high alkali content." According to the article by Kerman, some of the old men in Szechuan province said that their grandfathers, as boys, knew Li, and that he was then well along in years. Much of the secret of Li's long life is revealed in the statement that for the first century of his life he was an herb gatherer. In the pure, energizing air of the fields he laid the foundation that carried him through 256 years. Had he spent his early years grinding out his days in the sweat-shops of civilization, he had done well to live fifty years.

In his "Believe it or Not," Ripley stated that Numas De Cugna of Bengal, India, lived to be 370 years old. He grew four new sets of teeth, and his hair turned from black to grey four times, He died in 1566. Arphaxed, grandson of Noah, lived only 68 years longer than Cugna, dying at the age of 438 (Gen. 11:13). Dyson Carter stated in *The National Home Monthly* that scientists now assert it is well within the range of possibility for the average man to live 550 years.

According to the press of October 16th, 1941, Dr. Maurice Ernest, "one of the world's greatest authorities on longevity, said today," the account states, "that man can be made to live 200 to 300 years." He adds: "Many discoveries that point to the way of periodical rejuvenation have already been made."

Body Never More Than Seven Years Old

The body is incessantly renewing itself from the softest tissue to the hardest bone, and this process of renewal, according to physiologists, gives man a new body every seven years. In other words, the body is never more than seven years old no matter how many times the earth turns on its axis for a certain individual. No "periodical rejuvenation" is needed for such a body unless bad habits and bad environment have plunged it into degeneracy and decrepitude, as in the case of Captain Diamond.

Most centenarians on earth now live far from the polluted centers of civilization and industrialism. They are usually people of little means, of humble circumstances, who have been forced to lead a simple life and subsist on common, natural foods. Poverty is not the cause of sickness and short life, except insofar as it compels one to toil for a living in sweat-shops, filthy industrial plants, and stuffy offices filled with tobacco smoke and stagnant air. In the above list of very old people, no names of scientists and doctors appear. If they know how to live, their knowledge does them no good. Centenarians among scientists, doctors and the rich and opulent are rare, and when found it is discovered that they also live the simple life. Poverty enforces sobriety, frugality and the simple life of Nature. This course contents the body and prevents its vital channels from being clogged by exclave eating of denatured foods.

Professor Huxley fed worms as they usually eat, except one, which he fed the same, but occasionally fasted it. That worm was living and vigorous after nineteen generation, of its relatives had been born, lived their regular time, and died. If that were done in the case of man, he would live approximately 2000 years.

Fruit And Longevity

Herodotus wrote: "The oldest inhabitants of Greece, the Pelasgians, who came before the Dorian, Ionian and Eolian migrations, inhabited Arcadia and Thessaly, possessing the islands of Lesbos and Lokemanos, which were full of orange groves. The people, with their diet of dates and oranges, lived an average of more than 200 years."

Hesoid said: The Pelasgians and the peoples who came after them in Greece, ate Fruits of the virgin forests and blackberries from the fields." Plutarch observed: "The ancient Greeks, before the time of Lycurgus, ate nothing but fruits." Onomacritus of Athens, a contemporary of Peisistratos, said: "In the days before Lycurgus, each generation reached the age of 200 years."

Philochorus said of the Pelasgians: "Their heroic spirit and their strong arms to destroy their foe, were formed of shiny red apples from the forest. Apples were their favorite food, and the speed of their feet never lessened. They raced against stags and won. They lived for hundreds of years in the world of Cronus, but their vast stature never diminished as they grew old, even by a thumb's breadth. The dark lustre of their black hair was never tainted by a single silver thread. They lived so long they tired the winds of measuring Time, soaring above them."

What a blessing it would be for man if he could go back to those glorious days.

Doctors Do Not Live Long

No doctors are found in the above list of aged people. Their average life-span is short. Their medical training makes them so artificial that they know little about the natural life. The press of November 20th, 1941, reported that Dr. Richard C. Foster, President of the University of Alabama, "died last night of creeping paralysis" at the age of 46.

The Tampa Tribune of December 4th, 1945, stated that Dr. D. G. Meighan died in a Tampa Hospital after a long illness of six months, at the age of 47. For eleven years he was in charge of the U. S. Public Health Service in Tampa; was county physician there from 1926 to 1933, and before that was resident physician at the Gordon Keller Hospital. He was district surgeon for the A.C.L. Railroad for three years, and during World War II was acting surgeon for the U. S. Coast Guard Unit in Tampa.

The press of May 13th, 1952, stated that Dr. Jacob C. Kaplan, psychiatrist formerly with the Veterans Administration in Lexington, Kentucky, "died yesterday in Jewish Hospital." He was 54. Dr A. L. Bishop, age 57, professor of business administration at Yale since 1918, died May 8th, 1932, of "a heart attack." Dr. C. H. Ramelkamp, age 58, president of Illinois College since 1905, died April 5th, 1932, "after a long illness." Dr. Paul W. Horn, age 64, president of Texas Technological College, died April 13th, 1932, of "a heart attack."

Dr. J. R. Robertson, age 68, head of the history and political department of Bera College, died April 15th, 1932, cause not given. Dr. John Parmenter, age 70, one of the physicians who attended President McKinley after he was shot at Buffalo, died June 1st, 1932, cause not given.

The press of June 4th, 1944, reported that Dr. C. E. Ryan, age 60, nationally known physician, lecturer and writer on medical subjects, was "found dead in bed at his home about noon today." The press of June 4th, 1944, reported that Dr. C. E. Ryan, age 69, died of a heart attack as he was delivering a baby. Others stepped in to complete his duties."

The press of February 3rd, 1945, reported that Dr. Irving S. Cutter, age 69, medical director of Passavant Hospital, Dean Emeritus of the Northwestern University Medical School, and health columnist, died today after an illness of several weeks." For more than a decade Dr. Cutter, who knew how to live only 69 years, wrote "an informative column, 'How To Keep Well,' for the Chicago Tribune, and widely syndicated." He tried to teach others "how to keep well" while knowing so little about the Cosmic Science of Health that he died "after an illness of several weeks" at an age when the ignorant Indians of the hills and forests are still in their prime.

"The press of May lath, 1952, stated that Dr. Frank A. S. Kautz, "prominent Cincinnati obstetrician, died yesterday at Jewish Hospital after a brief illness." He was 76 and had practiced mediate for more than 50 years."

The press of January 8th, 1943, reported that "Dr. George W. Crile, famed 78-years-old surgeon-scientist, who believed that he performed the first direct blood transfusion, died today after receiving 25 of them in recent weeks." Medical doctors, who discover the folly of medicine and give up the use of drugs, vaccines and serums and turn to natural method of living, fare much better than regular orthodox doctors do, yet they are discredited by the medical organizations. Dr. J. H. Tilden gave up the practice of medicine, turned to Nature's way, and died September 1st 1940 "in his 90th year." Dr. John Harvey Kellogg of Battle Creek fame, who for 67 years never missed a monthly contribution to his Good Health journal which he edited all that time, and who ran a quarter of a mile each day, died in 1943 at the age of 91. He was a vegetarian for 76 years and seems not to have known that Vegetarianism is bad. He held that flesh carries too much contamination for safe consumption, and produces excessive intestinal putrefaction.

Live 200 To 300 Years

Colonel Robert McCarrison of the British Army Medical Staff, reported that during ten years' service in the Himalayan region he found no sickness of any sort in the colony of people where he was. He said: "Ages well beyond 250 years were common. Men of well attested ages up to 150 years were recently married and raising families of children. Men said to be well over 200 years of age were waking in the fields with younger men, doing as much work, and looking so much like the younger men, that I was not able to distinguish the old from the young."

There is no secret about the vigorous health and long life of these natives. They breathe to live, drink to live, and eat to live. In "20th Century Health Science" Dr. Francis X. Loughran said: "There are many immediate reasons why people die, but there is no underlying necessary reason that any scientist has yet discovered. In short, there is no principle limiting life."

Here is an extra item which came to our attention after finishing this lesson.... The Grit of January 20th, 1952, reports the case of J. R. Costello, age 87, of Winchester, Virginia, who had just finished cutting the seventh tooth of his third set "Dentists are mystified," adds the account.

Lesson No. 19
Water Causes Aging

"*If hardening of the arteries could be prevented, our life-span would be pushed far beyond the dreams of man.*" — Theo. R. Van Dellen, in his daily column "How To Keep Well" in the press of March 25th 1949. Van Dellen regarded as a mystery the cause of the hardening process, and admitted that "medical science" has no remedy for it. That made it another of the many so-called "incurable diseases."

In Lesson No. 16 we saw that vegetarianism is one cause of the hardened conditions that occur in the body. At the age of 79, after being a vegetarian for thirty years, Captain Diamond suffered from a serious state of ossification of tissues and blood-vessels and stiffness of muscles and joints. After given up to die by the doctors, Diamond turned for help to Nature, to the power that made him. He discarded vegetarianism, became a fruitarian, recovered health sufficiently to outlive all the doctors who gave him up as incurable, and reached the amazing age of 120. No one can say how long he might have lived had he been a fruitarian from the first. A group of eminent doctors made a careful study of the common process of sclerosis. They found that certain earthy minerals contained in tubers and cereals are one cause of the condition of ossification that produces decrepitude and premature death. One author states that the difference between youth and old age is a matter of chemical differences in the body, and not a question of years. He reported the case of a little girl of four years who developed a salt-eating habit to such extent, that in less than a year her entire body began to harden, her face to wrinkle, and she showed all the signs of Old Age. And man still eats salt. He also smokes when he knows it is a short cut to the grave.

The youthful body is supple, elastic, and vital; the aged body is stiff, rigid, creaky, sore, and achy. The condition of the body, not the passage of time, is the difference between a young man and an old man. It was demonstrated in the case of Thomas Parr, who lived 152 years, that it is possible to prevent ossification of the body. His case also demonstrated the dangers of eating freely of "nourishing food" as people are advised to do. This writer's maternal great grand-father was a vigorous man at the age of 110 when he died as the result of an accident.

These very old people "always die as the result of an accident" one author says. The law of averages overtakes them when they live so long.

The body of the infant is soft and pliant. The bones are plastic and flexible. After birth the bones begin to fill-in with mineral salts. This mineralization increases the size of the bones and their solidity. For this reason the growing child needs considerable lime, comparatively speaking, in the form of calcium automate and phosphate.

Less Minerals Needed After Maturity

The mineralization process is quite rapid until the body attains its growth. As growth stops, less lime salts are required for the body. This means that a change should be made in what one eats and drinks. When mineral salts enter the mature body in excess of requirements, they can no longer be used to develop and solidify the bones. Body development is done. So the excess minerals now begin to form damaging deposits in the body and its organs. Thus, the soft, pliant body of childhood becomes the hardened, stiffened body of old age. The spryness of youth becomes the slowness of decrepitude. Vitality decreases; senility comes. There is no mystery about the change. Dr. Logan Clendening wrote: "In youth the arteries are elastic, but as the body grows old, they become stiffer on account of the replacement of their elastic tissues by fibrous tissues and lime salts. In many cases the arteries may be markedly thickened and even so calcified as to have earned the term 'goose-neck arteries' because the deposit of lime salts give them a corrugated feeling like that of a goose as one feels its neck." — The Human Body.

It appears to be forgotten that the scleral process affects the entire body — cells, tissues, glands, blood-vessels. Some of the smallest blood-vessels become so hard and brittle that they burst under slight pressure.

Causes Of Sclerosis

The process of sclerosis, the condition of aging, rises from the following causes — (1) Bad air, (2) Bad water, and (3) Bad food.

The student knows what we mean by "bad food." He will soon know what we mean by "bad water." Yet the food and water we call bad are consumed by millions and considered good. Food is third in our list, yet most authors put it first. Then they give slight notice to the second, and none to the first and leading cause.

Water plays a big put in the ossification role because so much of the body is composed of water, and because most of the water used is the kind that produces ossification. The body of the average infant is approximately 75 percent water. A man of 150 pounds in weight would weigh 50 pound if he were well dried out. The more highly refined a tissue is, the greater the percent of water entering into its composition. The blood is 90 percent water; some parts of the nerve system are 90 percent water, others are 85 percent, and ordinary nerve tissue is about 80 percent water. The brain is 85 percent water, and the bones are nearly 50 percent water. People give great attention to what they eat, but little to the kind of water they drink. No matter how much city water is poisoned by political health boards, the people use it and offer no protest. It pays to keep people ignorant.

In all the fluids and tissues - blood, lymph, nerves, glands, muscles — water plays the part of a general solvent. By its work the River of Life is replenished and maintained. It is the medium in which solid and semi-solid aliment, are dissolved so they can pass into the blood, and by which all excretory products are eliminated from the body. The various processes of excretion, transudation, and elimination depend on water for their performance.

Importance Of Water

In the physiology of respiration, water and air unite, and water fulfills its biological function in the form of vapor mixed with the atmosphere. The importance of its needs in living organisms appears in the fact that water occupies four-fifths of the earth's surface. The dual elements of water and air in combination in the atmosphere assure the supply of the total needs of man's vitality. They do so in three principal ways. viz, 1. With the aid of the sun, which vaporizes the water of the earth, causing the water particles to unite with the atmospheric properties of the air so as to serve in the physiological process of respiration. 2. By hematosis, or the aeration of the blood in the lungs in the function of respiration, and 3. By supplying the body with the sustaining properties of oxygen and nitrogen in the air, and the oxygen and hydrogen in the water vapor.

To this list there must be added the emanations radiating from the sun's effluvia, which is condensed by the water vapor during its time in the atmosphere. This sun-filled vapor, inhaled by man, becomes an accumulator of the power of the sun within the body. Water deprived of sunlight it termed dead, and incapable of transfusing the least vitality into the organism. Such is the case with water from wells, closed cisterns and distilled water. Water should be exposed to the air and sunlight for sometime before used, so it may be regenerated with the vitality of the sun and air. But this does not apply to water contained in coconuts, melons, berries, fruits and vegetables.

The living cell is the basis and foundation of our body. The cell can function on the life level only in an aquiferous environment. If the cells are deprived of water, they become dry, inert and fall below the life plane in function. They are called dead. There are no signs of living things, plant or animal, in deserts where perpetual dryness of the air prevails. Living forms cannot come into physical being where there is no water.

Much care should be exercised as to the kind of water one takes into the body. One kind hardens blood vessels, tissues and glands, and produces stones in kidneys and gall bladder. Another kind of water dissolves and washes from the body the mineral deposits that produce these conditions. Most water used in general is what we call "hard." It is water in which quantities of lime and other minerals are held in solution. Spring water, well water, water that comes out of the ground, is charged with lime and minerals are held in solution. Hence, such water is "hard." Water from certain wells, springs, and lakes is often called "soft". It is soft only in comparison with water that is harder. Dr. G. A. Dorsey wrote: "Each year the earth's rivers carry to the sea billions of tons of dissolved minerals and carbon compounds."

"Mineral Water" Not Beneficial

Suffering people are often deceived by certain claims for "mineral water" from certain springs. Such water is very hard and very bad for the body. Those who recommend it for health are in error, or they profit on the sale of it. Some patients claim they get relief by drinking such water. They do at first, largely because they drink much of it and that aids in cleansing the glands and tissues. It is the second effect that is lasting and damaging. If the use of such water is continued, "serious effects will appear from mineral deposits in the body. So the remedy back-fires and in the end does much more harm than good.

Spring water, any water from the ground, contains in solution an amount of earthy ingredients that is fearful to contemplate. It has been calculated that water of the average quality from the ground contains so much carbonates and other compounds of lime, that one using the average quantity in the form of tea, coffee, soup, etc., would in forty years be sufficient to form a pillar of solid chalk or marble the size of a large man. So great is the amount of lime in spring and well water, that the quantity consumed daily would alone be sufficient to clog the system and bring on decrepitude and death before one reached the age of 20, but for the heroic labor of the eliminative organs. Were it not for the eliminative function of the skin and urinary system by which is eliminated much of the earthy matter entering the body with food and drink, no one, who eats and drinks in the conventional way, would live ten years.

Undistilled water, taken internally, is very bad. Boiling the water removed only a portion of the water and leaves the earthy matter beheld. In the hundredth part of a drop of raw water the microscope reveals a world of tiny animals. The dead bodies of the animals remain in the water after it is boiled, and help to clog the depurating organs and eliminating channels. By drinking boiled water one may avoid taking the live animals into the body, but one buries their dead remains in the body. If raw water is an aquarium, boiled water is cemetery. After boiling in a clean tea kettle for a week the water from a well, spring, creek, or lake, a stony coating will be found on the inside wall of the kettle.

Lime Deposits Cause Stiff Joints

What occurs in the body when one drinks such water for forty years? If the body were as helpless as the kettle to protect itself against the accumulation of these deposits, the body would be a solid pillar of limestone in a few years. Liquid lime is always present in the blood. When a structure or gland decomposes in the presence of liquid lime, the lime begins to fill the space resulting from the decomposition, and there it solidifies. Many fossil remains of ancient animals and vegetables are limestone casts; thus, filled in as the original entity decayed. Bunions and enlarged joints rise chiefly from this cause. When a joint is held open, as where an ill-fitting shoe holds the great toe pressed over towards the other toes, the liquid lime fills in the space thus caused, and the joint becomes enlarged. If any joint is held in one position long enough, without movement, it will grow stiff because of liquid lime deposited around the joint. The only remedy is movement of the joint and the use of distilled water, or the juice of oranges or grapefruit, to dissolve the lime so it may be washed out and eliminated.

Rain Water And Distilled Water Are Safe

Rain water has been distilled by the sun. It is free of all minerals. But when it falls as rain, it may pass through air filled with tiny animals, dust, smoke, soot, acids and all kinds of filth. As it reaches the earth in such cases it is so saturated with the filth of civilization, that its color is a very light straw. As the rain continues to fall, the air soon becomes washed of the filth, and the water grows clear and clean. That is the water one should use. Distilled water is water that has been transformed to vapor and condensed. It is free of minerals and the only water, except rain water, fit to use. Such water may be boiled in the same kettle for years and will leave no deposits on the kettle's walls.

Distilled water is the greatest solvent known. It is the only water, except clean rain water, that may be taken into the body without damage to cells and tissues. By its continued use, it is possible to dissolve mineral deposits, acid crystals, and other hardening deposits in the body. Captain Diamond, mentioned in Lesson. No. 16, got relief from his stiffened state by the use of fresh fruit juices and distilled water. As these are not "medicines," they are not usually used.

As distilled water is such a powerful solvent, it is bad for the teeth because it leeches out their minerals. The same is true, to a lesser extent, of acid juices such as those of oranges and grapefruit. Distilled water passes directly into the blood and the solvent properties, of the blood are increased by the distilled water to a degree that the blood will keep in solution the mineral salts already in it, and prevent their harmful deposition in organs and glands, and favor their elimination by the different excreta. If distilled water be taken in large quantities, or if it be the only liquid one takes into the body, it will in time tend to dissolve and remove those earthy compounds that have accumulated in the system, the effects of which usually become more manifest at the ages of 40 and 50. The daily use of distilled water facilitates the removal of deleterious compounds from the body by means of the excreta, and therefore tends to prolong life. No water is as good for the body as the distilled water contained in coconuts, melons, berries and fruits. From this source one should obtain all the fluid the body needs, and plain distilled water should not be used unless one believes it is necessary to dissolve tumors and hardening in the body.

People living in limestone localities who use water from wells, springs, creeks, lakes, are invariably afflicted at a comparatively early age with a general ossification of the whole body. Instances of longevity among such are rare. As we eat to live and eat to die, so we drink to live and drink to die.

Lesson No. 20
The Wonderful Orange
by
Dr. Leon A. Wilcox

It was Kipling who said:

*"If you can bear to hear the truth you've spoken,
Twisted by knaves to make a trap for fools,
You'll be a man my son."*

For twenty years I have been trying to put over a message about the wonderful orange and the benefits to be obtained from its use. I have heard the truth, I have spoken twisted, pulled apart, distorted and tortured, still, had it not been for the great truth it is, there would not be enough of its virtues left to wad a pop gun.

Until a very few years ago, it was the custom to carry home a dozen oranges with much the same mental attitude as was felt about a box of candy. They had been regarded as something nice to eat between meals, or as a dessert. To think of them as a wonderful food — perish the thought. The majority of the medical professions have always knocked citrus fruits. Even in this day of enlightened opinions from some of the world's most famous physicians and dietitians, it is not un-common to hear one say their doctor told them not to eat fruits, as they contained much acid. Especially are people advised not to eat the citrus fruits, such as oranges, lemons and grapefruit, on account of the acid. It is only within recent years that any of the medical practitioners suggested the use of these valuable foods. I regret to have to say that only a few today know now these fruits should be used. It depends entirely how these fruits are used as to whether they will give an acid reaction or not. Citrus fruits will always return an alkaline reaction when taken into the stomach by themselves. These fruits (in fact all juicy fruits) should never be eaten at the same meal with cooked or baked foods, nor should they ever be eaten with sugar.

Cooked foods contain a certain amount of starch. When fruit juices come in contact with starch in the stomach, the reaction is certain to be fermentation. The fermenting process is what generates acid. So you see it is not the fruits that make the acid, it is the food combination.

A common sight in any restaurant in the morning, at breakfast, is people drinking a glass of orange juice or eating grapefruit, followed by a sweet roll, then washed down with a cup of coffee. This makes a nice acid breakfast and, if continued long enough, will produce an acid stomach, neuritis or some kind of rheumatic condition.

Remember, all kinds of fresh fruits, melons, and berries should always be eaten alone, or with the fresh salad vegetables. William H. Dieffenbach, M. D., of New York City, is authority for the following about fruits: "Fruit instead of being a dessert, should, if properly evaluated, be classified as the most valuable of foods. Fruits contain little protein and fat but are most valuable sources of mineral salts, cellulose, carbohydrates, and vitamins. The water content of fruits, with mineral content, keeps the blood in a state of alkalinity. Its alkaline elements, which are combined with the fruit acids, act as natural laxatives by promoting the secretory action of the liver, pancreas and other secretory glands. Fruits furnish calcium, potassium, magnesium, phosphorus, iron, and manganese of a highly organized type and are indispensable for the re-building of red blood platelets and corpuscles. The fresh citrus fruits, lemon, orange, tangerine, lime and grapefruit, prevent scurvy, due to vitamin C."

King Of Fruits

After having given the question of fruits twenty years of study and research not only by personal use but also by the direction and treatment of patients, I have arrived at the conclusion that the orange is king of all fruits. Very few would believe the length of time an individual can live and perform the hardest kind of work, both mental and physical, using absolutely nothing but oranges for food. Another very important thing: a sick person living exclusively on an orange diet, is not only getting all the nourishment that the body requires, but the orange will neutralize the acid in the system.

Finest Distilled Water

I, one time, heard a great physician and dietitian say, "orange juice is water distilled in God's own distillery." How true this is. It is a food and a drink for the healthy and medicine foe the sick. For those who are seeking to regain health, there is nothing that can be taken that will assist nature like this golden elixir of life.

Millions of dollars are being spent annually advertising all kinds of foods. The old high powered salesmen have given way to the high powered advertising counselor. These fellows must lay awake nights conjuring up good advertising copy telling us why we should eat Buncom & Cos Patent Leather Cheese for health. Having many friends and relatives among the advertising fraternity, printers and allied lines of business, I am much amused, at times, to hear remarks which indicate how easy it is for these high pitch copy writers even to put it over on their own profession. Well, there is nothing like taking your own medicine. For one advertising man to believe what another fellow worker says about the product he boosts surely shows faith in the profession. However, I would suggest that in the future you do not place much dependence on what advertisers have to say about their food products.

In the fruits and green uncooked vegetable; you have food exactly as Nature has prepared it, and there is positively nothing of any sort that can equal these foods for health. All the vitamin content is still intact. There has been no processing of any sort to remove the virtue from these things. All the food value is still there. The pioneers brought to us the beautiful California and Florida sunshine and all its wonderful fruits. I am proud to be one of the pioneers who have been teaching people to use these products of the climate and sunshine discovered by some of our forefathers. Let us learn to utilize the golden nuggets of health found in fruits — (Typo Graphic, Pittsburgh, Pennsylvania, February 1981) — Pittsburgh Health Club.

Now we present a most interesting account written by John W. Marshall about a person who lived six months on orange juice. This person was a patient of the author of the above article on "The Wonderful Orange."

Six Months On Orange Juice
By
John W. Marshall

For years I had known of the great food value of the orange, not its value as measured in calories, but as a rectifier of the chemistry of the blood, as a restorer of the proper alkalinity of the life stream. I had seen many people live two or three weeks and even a month and on occasion even six weeks, on an exclusive diet of the delicious fruit. I had seen people complaining of all sorts of diseases, afflicted with worn out, poisoned, overfed and over nourished bodies, restored to a remarkable degree of health through the exclusive use of the citric fruit for varied periods of time. But when Dr. Leon A. Wilcox, a leading Osteopathic Physician of Pittsburgh, Pennsylvania, informed me in an unassuming fashion that he had a patient who had lived on orange juice for six months, I was amazed. I should not have believed the story from the lips of an ordinary man. But Dr. Wilcox is a man of high repute among the members of his profession and among his many patients and friends in Pittsburgh, where he has lived and practiced for a quarter of a century. Then, too, when he told me the story as I sat at his office, he spoke with such confidence and candor that I never thought for a moment to doubt his word. The following day I had the pleasure of seeing and speaking with the patient herself, a quiet little girl with big blue eyes and an oval face encased in a lovely white skin, into which, as I conversed with her, there came and went flushes of pink and red as she told me the remarkable story of her recovery from a condition of living death and her gradual evolution into a creature of living delight unto herself and to all whom she might meet.

This is the story she told me on the November evening as we sat in Dr. Wilcox's office: "When I was a little girl, I was just about like other girls. I had fair health most of the time, though I was visited by the usual so-called children's diseases. I suppose you had a cold occasionally," I interrupted. Of course, plenty of them, especially in the winter time," was the reply. Then she continued: "At about fourteen I began to get fat. My parents, the stocky German type, heralded this acquisition of weight as a sign of health. Of course, I thought the same, as I did feel quite well most of the time. However, I had a voracious appetite and ate not only prodigious quantities of the 'good staple' foods, such as bread, meat and potatoes, but great quantities of candies, ice cream, etc., etc. Certainly the most iron-bound constitution must have given way under the load.

"As the time went on I got fatter and fatter and my complexion, once ruddy and beautiful, began to acquire a sallow yellowish appearance. Blackheads and pimples became numerous. To rid myself of the latter I tried various lotions, cold creams, beauty clays, etc. To restore the roses in my cheeks I tried various highly perfumed toilet soaps and, of course, rouge and powder. When I did not succeed in eliminating the pimples and blackheads by the use of the skin lotions, and when beauty soaps failed to restore the roses to my cheeks I used more and more rouge, lipstick and powder. Of course, it never occurred to me that my voracious and unbridled appetite had anything to do with my complexion, though my weight was steadily increasing. Dark rings began to appear beneath my eyes. I began to have headaches, which as time went on became more frequent. At first I sought relief in aspirin tablets, which of course relieved my headaches, but I realized that my condition was growing worse so I began to visit doctors, from whom I got pills and prescriptions and orders to have my teeth pulled and my tonsils removed and various suggestions of equally stupid character. Of course I did not realize then that all these things were stupid, though it is true I kept both my tonsils and teeth.

"If the doctors had been unanimous in their analyses of case and given the same prescriptions I should probably have followed all of the advice given, but the prescriptions varied so much that it was impossible to follow them. My appetite began to wane. My tongue was heavily coated, especially when I got up in the morning. But I took appetizers and ate highly spiced foods so that I was able to eat in spite of my revolting stomach. My headaches increased and my pains extended to other parts of the body. My legs, my arms and especially my back ached most of the time, As I was now employed with the Westinghouse Electric Company and wanted to be always on the job, I had to force myself to do my duties; force myself to get up in the morning; force myself into my clothes; in fact, force every move that I made. In the end, even eating became a burden to me. The only act I did not force was going to bed at night, but my sleep was never sound and dreamless. Instead I rolled and tossed all night with occasional lapses of consciousness. Whenever I rode on the train and occasionally even at work, I drowsed off into a stupor.

"Not only because of my suffering and unsightly condition, but because I discovered that while other girls were in demand, I remained a wall flower, I stayed away from dances and social gatherings of young people. I became exceedingly morose and morbid and more and more self-centered. Life had become such a burden to me that many times, in moments of greatest depression, I contemplated suicide, and only refrained because I lacked the courage. When doctors' medicines failed, the neighbors advised herb teas, mud bath packs, grandmother's physic, etc., but these home remedies were, no more effective than those prescribed by the doctors. Instead of getting better, I got worse. "My heart which had been for some time troubling me a great deal, at times thumped so rapidly that it seemed it would jump out of my throat. My breath became short. My pains increased. My flesh became soft and pudgy. My ankles became almost as large as my calves. I was a sight to behold — only one and a half inches over five feet tall, barefoot, and weighing one hundred and eighty-six pounds.

"Here I was, only eighteen years of age, as big as a baby elephant and saturated with the poisonous wastes the food I had been eating. As I was about to give up in despair I heard of Dr. Wilcox, 'the man who cures people with oranges.' So, without a great deal of hope, but with the feeling that regardless of whether or not he could benefit me, since my condition was so bad that he could not possibly make me worse, I went to him, and for the first time since I had begun doctoring, I was told the real cause of my trouble. I was told that neither my stomach, nor my heart, nor my under-nourished, decaying teeth, but my diet was responsible for my trouble.

"After a thorough examination, the doctor said, 'We'll just put you on orange juice for ten days as a starter on the cleansing process.' Of course this seemed like a long time to do without what I called food, but I was desperate, so I said, 'All right. I'll do anything. I might as well be dead as in my present condition.' The results were surprising, not only to me, but even to Dr. Wilcox, who had witnessed so many people doing the same thing. While the first few days were a bit difficult. I began to experience immediate relief, and before the ten days were up I had lost all of my pains and I have never had a headache since. When the ten days were finished, I felt so much better that I decided, upon the doctor's advice, to try the same diet for ten days more. This was not hard to do, for I felt no desire for other food. At the end of twenty days I felt still better and my fat was rapidly dropping away, so as I still had no desire for other food, I continued on the exclusive orange juice diet. So I went on from one ten-day period to another, and as the days passed into weeks and the weeks into months, as my desire for other foods had not returned, as I had long ago said goodbye to my pains, as my fat was melting away, and my complexion clearing up, I continued my course.

"Life had taken on a new meaning. I had begun to enjoy living. I no longer drowsed on every occasion when I relaxed. But when I went to bed, my sleep was sound and untroubled, in contrast to the spasmodic sleep that gave me no rest in the days when I was living in the old way. I became active and alert, full of vigor and vitality. Boys and girls alike began to desire my company, and the former especially became increasingly interested in my new found charms. So I continued day after day and week after week until now six months have passed and I am, as you see, completely restored to health. And I want to tell you it is great to be alive. I feel like running and dancing and singing all the time."

To see was to believe. I compared the pictures she showed me of the overfed, overstuffed creature she had been, with the living, breathing reality before me, and I knew her story was true. "You seem to have gone through this period purging without pain. This is unusual. People, as a rule, suffer somewhat, especially at the outset of such a restrictive diet. Did you not at times suffer and feel morbid and discouraged?" I asked. "No", she replied, "I improved from the start, and although the first few days were painful, I felt better each day. At about the middle of the period I had a slight running at the nose (Dr. Wilcox called it a period of elimination), but this did not bother me I kept on as usual with my work in the office of the Westinghouse Electric Company? Then you worked throughout the period of six months that you were on the orange juice diet?" I again interrogated. Yes, and I walked about a mile every day and felt like walking more, but the doctor cautioned against it."

"You seem to have kept in pretty close touch with Dr. Wilcox throughout the period? Yes, I visited him every day. While he emphasized the fact that not he but the oranges were doing the work, he felt that without his guidance I might go wrong. I probably would have, too," she added. What do you eat now?" I asked. "Raw foods, altogether, green vegetables and a few nuts. This diet I enjoy much better than did I the old conventional cooked diet," she added with a smile of conviction.

And thus ended the story of the wonderful transformation wrought by the daily use of the golden drops of sunshine from the orange. May it be told again and again. May it be an inspiration to thousands of suffering human beings, that they may be tempted to partake freely of this golden fruit whose substance has imbibed so freely of the life-giving properties of the sun that even the color of its skin bespeaks the gold that lies within. — Correct Eating 1931. Read over carefully about this girl who began to fatten when she was 14, and how her parents considered it a sign of good health. Then her health began to fade... appetite began to wane, tongue heavily coated... headache increased and pains extended to other parts... heart went bad, breath became short, flesh soft and pudgy, ankles almost as large as calves of her legs. She was 18 years old, 5'1 ½" tall, and weighed 186 pounds; ready to give up in despair when Dr. Wilcox put her on orange juice. Improvement was rapid and surprising. Within ten days she felt much better and continued the orange juice for six months – and was *"completely restored to health"*. While on the orange juice diet she carried on her regular work "in the office of the Westinghouse Electric Company". Then she changed to a diet of "raw foods entirely, green vegetables and a few nuts". She said, "This diet I enjoy much more than I did the old conventional cooked diet". This young woman made an excellent start on the path to Breatharianism, and no doubt had continued had she been properly advised. But no one knows anything about Breatharianism and the advice she needed no one could give.

The turning of the earth on its axis affects not the body. The fault is not in our body but in our conduct and habits if we grow decrepit. The press of May 3rd, 1936, reported the case of a woman who neither ate nor drank for 58 years, and — *"at the age of 68 she acts and looks like a child."*

Lesson No. 21
Breath Of Life

"The Essence of the Universe is in the Infinite Air in eternal movement which contains ALL in itself. Everything is formed by integration and disintegration of the AIR under the Law of Expansion and Contraction." — Anaxmens

We saw in the preceding lessons that man eats to die and drinks to die. Now we shall learn that he breathes to die. Breathing is such an easy, natural function that people give it little attention and regard it lightly. Until recent discoveries in the field of atoms, only a few realized that the Essence of the Universe is in the Infinite Air in eternal movement which contains ALL in itself. Man has gone without eating for weeks and lived. It is reported that some people have lived for years without eating. Man has gone without drinking for 30 days and lived. But if he stops breathing for three or four minutes, it is fatal. This is proof that breathing is the big secret of living. *When we stop the Breath we stop the Life.*

According to ancient wisdom, the mystery of Life is not in the body's function, nor in food, nor in chemical changes occurring within the body, nor in the decomposition of the body's tissues-but in the "spiritus Nitro-aerius." The Breath is the Life (Gen. 2:7). The Spirit of Life animates the body; the flesh profits nothing, remaining a chemical compound of atoms. When the Breath of Life no longer animates the body, it disintegrates and its atomic elements return to their original source (Eccl. 11:2; Jn. 6:63). That is the philosophy of the Ancient Masters. It sounds sensible and its truth is hourly proven. Had this ancient secret of Life been lost, the world had nothing to guide it in this important field but the absurd theory of modern science, that Life is the expression of a series of chemical changes (Osler). When we recognize that "Breath Is Life," as Pandit Acharya puts it, we have a definite law of biology, psychology and physiology — a law that modern science up to this hour has not.

Early Theories Of Respiration

The Secret of Life was confirmed by the Masters to their Mystery Schools, and imparted only to the Initiates. It was unknown to the masses. That accounts for the stupid theories of Respiration that prevailed in the days of Aristotle (384-321 BC.). From his day down to the 15th century A.D., it was believed by science that the purpose of breathing was *"to draw air into the body to cool the blood."* Out of this theory came the absurd Galenic doctrine (131-210 A.D) that: "Air introduced into the body by breathing saved to regulate to maintain and at the same time to temper, to refrigerate the innate heat of the heart."

It is shocking to learn how little was known about Respiration by the supposedly intelligent men who laid the foundation of modern science. It was chiefly through the work of a group of English men in the 17th century that occurred the unraveling of some of the secrets of Respiration. In 1667 the discovery was made that air is absolutely essential to the life of animals and that the gases of the inhaled air enter into and become part of the blood.

What were men of science doing that less than three centuries ago they knew not that Air is positively essential to the life of animals? At that time chemical knowledge was so deficient that nothing was known of what occurred after the inhaled air entered into and mixed with the blood.

That secret was explained in the ancient records, some fragments of which Constantine's army failed to destroy in the 4th century, after it had been decreed at the First Council of Nicea that the Ancient Wisdom must be destroyed, as a result of which destruction of the Roman Empire and all in provinces were plunged into a reign of darkness that ruled for a thousand years. Gibbon did not write a true story of the "Decline and Fall" of Rome. The next step in the dark realm of Life was the discovery that the difference between the dark, venous blood, and the bright red arterial blood is due to the admixture of gases in the air. Until this discovery, modern science had considered Air as a very simple substance and not a complex compound. It had sneered the statement that *"the Essence of the Universe is in the Infinite Air in eternal movement which contains ALL In itself."*

Physico-Chemical Theory Of Respiration Rediscovered

It remained for John Mayow, in 1643, to discover what was well known to ancient science. All the air inhaled is not used by the lungs to influence the blood, but only a certain part, which he called "Spiritus Nitro-aerius," and which was later termed "oxygen." Mayow thus discovered part of the ancient secret of animation, and developed the first faint physico-chemical theory of Respiration in Modern times. He said: "With respect to the uses of respiration, it may be affirmed that an aerial something, whatever it may be, essential to life, passes into the blood (from the air). Thus the air expelled by the lungs, these vital particles having been extracted from it, is no longer fit to breathe again."

This remarkable discovery meant so little to medical art, that Mayow's work lay neglected and forgotten for almost a hundred years. The secret of animation, of Life, was exposed before their eyes, yet they saw it not. In 1774 Priestly re-discovered Mayow's "Spiritus Nitro-aerius" and isolated a gas he termed oxygen. But it remained for Lavisier (1782) to show what oxygen is, thus throwing more light on Respiration, but failing to find the secret of animation. It was not until the middle of the 19th century, when Gustay Magnus proved the presence of the blood-gases, in different proportions in the blood, that the modem theory of Respiration assumed anything like definite form. Almost another century was destined to elapse before medical art considered Air sufficiently important in relation to health and life to make a special study of it.

In 1924 a group of physicians working at the St. Louis Infirmary in cooperation with Washington University, concluded from their study of 1000 persons, that better health and longer life for middle-aged people may be achieved by *"maintaining the proper level of oxygen consumption in the body."* The group found that the ideal oxygen consumption occurs in the first ten years of life, when the lungs are in good condition and chest expansion is greatest. Then the rate of oxygen consumption declines—but the reason why was not stated. By the time a child is ten, the amount of oxygen consumption begins to decline because shallow breathing begins, because of lung degeneration, because of the polluted air of civilization. The epoch-making discovery that Air is of paramount importance in matters of Life and Health arrived too late. Text-books had already been filled with the theories to the effect that Life is *"The expression of a series of chemical changes"* – (Osler), that air has little or nothing to do with the matter.

Until about fifty years ago, it was considered by the doctors as being so dangerous to the sick, that when the medical doctor called, after an examination of the patient, he ordered windows closed and fastened down, and all cracks and air-holes plugged with cotton to keep out the air. He further ordered a heavy blanket hung round the bed so as little air as possible could reach the patient. It took Dr. Bremer of Germany sixty years to convince people that air is good for the sick. After he forced the doctors of this country to recognize the truth, the medical association sent one of its leading lights into the New York mountains on a "fishing trip." This doctor "discovered" that outside air is not injurious to the sick, but actually beneficial.

Breathing Primary Function

We stated in Lesson No. 12 that while eating and drinking are voluntary and controlled practices, Respiration is an automatic, involuntary process, so far beyond man's conscious control, that he breathes when unconscious in sleep, or from injury, even better and deeper, more regularly and rhythmically, than when conscious and awake. Respiration is not only automatic and involuntary, but the primary function of the living organism. All other functions are secondary and designed to keep the body fit to perform the breathing function.

The lungs are definitely designed for and adapted to their work. They are by far the largest organs in the body, filling the thorax from the collar bone to the lower-most ribs, and from the sternum in front to the spine in back. The Lungs are truly the Organs of Life. When you stop breathing you stop living; and when you die, you go gasping for breath. Professor J. S. Haldane of England, in his work on Respiration, wrote: "Living is actually a struggle for air. Keep the vast lung surface of the body supplied with fresh air, and observe all other health rules, and there is, speaking scientifically, no known reason why you should ever die."

The living Organism contacts the realms of Spiritual and Material Substance through the respiratory and digestive organs. The body is equipped with Dual Centers to insure: (1) its preservation and (2) its perpetuation. The first of these centers are the Organs of Respiration. They are for the preservation of the organism itself. The second are the Organs of Generation. They are for the perpetuation of the race. The Breathing Centers, the Spiritual Organs, are the point in physiology where the ignorance of physical science "is profound" to use the words of the great Carrel. He wrote: "In fact, our ignorance (of the body and its function) is profound. Our knowledge of man is still most rudimentary.

Our knowledge of the human body is, in truth, most rudimentary. It is impossible to grasps its constitution. An endocrinologist, a psychoanalyst, a biological chemist are equally ignorant of man. Our knowledge of man is still rudimentary" (Man The Unknown, Pp. 4, 5, 109, 289).

Dr. A. E. Crews, professor of Edinburgh University, wrote: "It is more difficult to explain why man dies than why he does not live forever."

Why Man Dies

If we know not why man dies, it might be well to learn why he lives, as such knowledge may disclose the reason of his death. "And He breathed into them the Breath of Life. Life came with the Breath, and so Breath became the Principle of Life. But it goes deeper than that. We must breathe to get food from the air with which to sustain life. Breathing properly is the first requisite to health" (Editor of Nature's Path).

Some may sneer at that statement and contend that there is no Life Principle, that vital function is the result of a series of chemical changes (Osler). Ancient science held that through the nose and mouth, through his respiratory and digestive organs, the Spiritual Essence of the Universe enters man's organism, thus linking him directly with God. The absolute and uninterrupted persistence of the function of respiration is the leading wonder of the living organism.

As the new-born baby takes its first inhalation, it begins that process of breathing which will never cease except in death. Its cessation for only a few minutes is fatal. This knowledge discloses why man lives and why he dies. It narrows the search for the secret why man dies to the point where we must discover why respiration fails. The process of breathing contacts man with God. Were that contact never disrupted, it would mean Eternal Physical Existence and Eternal Knowledge, according to Herbert Spencer (1820-1903).

Why Man Lives

Spencer gave the Modern world the most scientific explanation why man lives and dies. He wrote: "Perfect correspondence would be

Perfect Life. Were there no changes in the Environment but such as the Organism had adapted changes to meet, and were it never to fail in the efficiency with which it met them, there would be Eternal (Physical) Existence and Eternal Knowledge." (Principles of Biology).

Spencer committed the error of believing that God's work is imperfect, and that man's somatic demise results from the faulty construction of his organism. If we can define scientifically a condition in which the organism would continue to live without end, then eternal physical life becomes a reality attainable by supplying that condition.

Professor Weissman said that: "Death is not a primitive attribute of living matter, but is of secondary origin." Dr. James T. Monroe supported that assertion when he wrote: "The human frame as a machine is perfect. It contains within itself no marks by which we can possibly predict its decay. It is apparently intended to go on forever" (Advanced Physiology).

Not only is man a mystery, but equally mysterious is the reason why he dies. Dr. William Hammond, late Surgeon General, U. S. Army, wrote: "There is no physiological reason known at the present day why man should die" (How To Live Forever).

Biology shows that man could not come into physical being until the condition of his physical Environment met the requirements of the Law of Perfect Correspondence. We know that condition prevailed when the first man came into physical existence. We feel safe in assuming that these harmonic conditions continued so closely to perfection, that early man lived for thousands of years. God's Plan of Life intended that Perfect Correspondence should always prevail as between man and his environment. That would fulfill Spencer's Law of Eternal Physical Existence and Eternal Knowledge.

It appears logical that man's death was due to his failure to live in complete harmony with the law that made him a physical being. Spencer cannot be wrong in holding that perfect correspondence would be perfect life. That is the Law of Physical Existence. But he cannot be right in assuming that the discord rises from faulty construction and constitution of the body. In that case man could not have come into physical being. The biblical record shows that it was Adam's misconduct, not the faulty constitution of his body, that created the discordant condition that caused his somatic demise (Gen. 2, 3, 4).

Professor Henry Drummond supported the theory of man's contact with God when he referred to the "scientific principle of continuity existing from the physical world to the spiritual." While ancient records and later discoveries support Drummond's assertion, it is so directly opposed to the theory of Evolution that physical science rejects it, and suppresses all fact which upset its theories.

While the line of contact between man and God remains normal, perfect correspondence prevails. No reason for its disruption appears except that the contact is weakened by the degeneracy of man's body, due to the artificial environment he has built for himself, and the faulty habits he has acquired. Civilized man has long been a stranger to the Natural Environment in which the race first came into physical being. Carrel says: "The environment which moulded the body and soul of our ancestors during many millenniums has not been replaced by another" (Man The Unknown P. 10). The constitution of man's body has not changed to meet the new conditions of his artificial environment that has replaced his natural one. The result is that of perpetual discord between man and his environment. The effect of this discord is a general deterioration of man's body, the symptoms of which are termed disease. Man's artificial environment and faulty habits cause degeneration to progress, steadily in his body, causing the line of contact with God to weaken. The result is a decrease in the Breath of Life that leads to somatic death.

It is similar to a machine that weakens and fails because of faulty contact with its source of power. The electric machine stops and the electric light dies when contact with the source of power is disrupted or severed. Man is governed by a similar law. He weakens and dies when contact with his source of power is disrupted or severed. His physical demise results because for him perfect correspondence with the Conditions of Life no longer prevails.

Lesson No. 22
Spiritual Organs

"Each part of the body seems to know the present and future needs of the whole, and acts accordingly. The significance of Time and Space is not the same for our Cells as for our (physical) Mind. The body perceives the remote as well as the near, the future as well as the present" (Carrel, in Man The Unknown, P. 197). We saw how amazingly little modern science knows of the Breath of Life and the function of Respiration. It has not yet discovered in the body the Spiritual Organs of man. Ancient writings are filled with references to the Spiritual World. Modern science holds that such world is a myth — that all is material substance and mechanical energy. Ancient science taught that man is a miniature Universe (Microcosm). Hence, if there is a Spiritual Realm in the Universe (Macrocosm), there must also be one in man. If that be true, man's body must contain organs through which the Spiritual Realm may manifest itself on the material plane.

Not long ago modern science regarded the air as empty and void. Thousands of years before the Masters taught that "the Essence of the Universe is in the Infinite Air in eternal movement which contains ALL in itself." All animals exhibited strange powers. Hornets and wasps have always known how to make paper. They were never taught and needed no experience. Whence comes this knowledge? Birds have always built their nests as they do now, and each kind builds a certain type of nest. They were never taught, and needed no experience. Whence comes this knowledge? Birds know which way to travel and when to avoid winter's icy blast. They know that snow and ice will come at a certain time, and they must fly in a definite direction to a certain region to escape the fate of being frozen to death. Whence comes this knowledge?

Modern science has no rational answer for these questions. The best it can do is to suggest that the birds and beasts are guided by "instinct." It fails to explain what "instinct" is, and assumes that it must be a property of Matter.

Materialism Is A Superstition

Modern physics has studied phenomena in matter around us. That brand of physics died with the discovery of the electron. Physicists are now busy trying to make the electron fit their materialism. They refuse to understand that the electron belongs to another world — the Spiritual World of the Masters. Professor J. S. Haldane, noted English astronomer, said: "Materialism, once a plausible theory, is now the fatalistic creed of thousands (of modern scientists), but materialism is nothing better than a superstitution, on the same level as a belief in witches and devils. The materialist theory (of science) is bankrupt."

And so is the materialist theory of Evolution. Radio, Radar and Television are mechanized examples of the Spiritual Powers that operate as Vital Intelligence in the strange conduct of birds and beasts, which science calls Instinct. Why does man not have these powers? The Ancient Masters taught that there is a Spiritual Realm in man. "The kingdom of God is within you" (Lu. 17:21). That includes everything and all. The Spiritual Realm of God in man is located in the Spiritual Chambers of the skull, called the Golden Bowl by the Masters (Eccl. 12:6). These Chambers, the function of which is unknown to modern science, are Five in number. The Masters called them the Five Stars of the Microcosm, and they are symbolized in ancient scriptures by certain fives, as the Five Golden Emerods (1 S. 6:4); the Five Loaves (Mat. 14:17), etc. The Sankhya doctrine states that the Five Physical Senses of conscious man are the exteriorized products of the five corresponding Spiritual Centers, which are as follows: FRONTAL SINUS — A cavity in the frontal bone of the skull. 1. SPHENOIDAL SINUS — A cavity in the sphenoid bone of the skull. 2. MAXILLARY SINUS — Largest of the five, and resembles a pyramid in shape. 3. PALATINE SINUS — A cavity in the orbital process of the palatine bone and opening into either the sphenoidal or a posterior ethmoidal sinus. 4. ETHMOIDAL SINUS — This chamber consists of numerous small cavities occupying the labyrinth of the ethmoid bone, and in these cavities are situated the small, mysterious glands known in Occult Science as the Intellectual Organs.

The Sinuses communicate directly or indirectly with the nasal cavity; and it is highly significant to observe that they receive the Breath of Life directly and unmodified as it flows from the Universe to them through the nose, and before any of the other air organs have a chance to select and absorb any substance from the Spiritual Essence of the cosmos, charged with every known and unknown element. The Sinuses are lined with the mucus membrane extending into them from the nose, and to them rapidly spreads all disorders that affect the nose. They receive without protection the full charge of all poisonous gases and acids in the air.

The nose is the first organ that reacts to polluted air, and that reaction is called a "cold." The inflammation resulting from the effect of the polluted air extends from the nasal mucus lining to that of the Sinuses, causing such disorders as frontal headache (frontal sinus), pain in the cheek (maxillary sinus), pain between the eyes (ethmoidal sinuses), and deep seated pain at back of eyes (sphenoidal sinus). These aches and pains, indicating serious damage being done to these Spiritual Chambers, are caused by poisonous air entering the nostrils. The air may be so slightly polluted that it fails to produce the reaction called the "simple cold." Thus, begins the destruction of the vital Spiritual Centers of man while he is only an infant — and the truth becomes known, that "cold" is a sign of serious damage being done, and not so "simple" as some think.

The mucus excretions of the lining of the maxillary sinus, in inflammatory conditions, fill up the sinus, as the orifice is at the uppermost part. Much of the mucus as cannot be blown out through the nose, remains in the sinus where it gradually hardens, destroying the spiritual function of that chamber — the largest of the group.

Recovery From Illness Only Partial

Full recovery from ailments is a myth. Each one is a step down the ladder of degeneration to the grave at the bottom. If the illness is slight, the downward step is short. If severe, the downward step is longer. Recoveries from each illness are only partial, regardless of how slight the illness may be. But if degeneration has not gone too far, a change in one's mode of living that brings the body in harmony with God's law of Life, will result in Regeneration.

The sinuses superficially appear to some as nothing more than air chambers in the skull. They are ignorant of their true function, and assume that their purpose is to lend resonance to the voice. The voice organ is in the throat, not in the nose or in the sinuses. Occult Science, termed by modern science as "that school of stupid superstition." teaches that in these Spiritual Chambers is located the seat of the Intellectual Divinity of man. These air chambers and the small glands in them constitute the spiritual sense-centers that receive from the Cosmic Source the Higher Intelligence which is too subtle for contact by the five physical senses of conscious man, in his present degenerate state. But this was not so when man enjoyed full Physical Perfection. Into these chambers these incessantly flows from the Cosmic Ocean of Ether a peculiar gaseous substance, a subtle essence, known to the Ancient Masters as Mental Spirit. It can produce no normal reaction in the spiritual chambers of civilized man, as they are deficient dormant degenerated by the evil work of polluted air. The small glands, the Intellectual Organs, located in the skull near the point where the nose pins the forehead, are activated by the Mental Spirit that passes through the nostrils into the sinuses, and, when they are normal and functional, they coordinate and collaborate with the sinuses. This is the chief Spiritual Intelligent Center of man.

The Kingdom Of God Within

In wild birds and beasts, and the wild natives that have not been tainted and tinged by the "blessings" of civilization, these spiritual centers are functionally developed – and modern science attempts to explain the uncanny powers of these creatures by asserting that they are guided by "instinct" but we are not told what "instinct" is.

If a hunting dog be kept in the house and breathe the polluted air the same as the members of the family, in time the nerves in the nose and sinuses become dull, the dog loss its keen sense of smell and is unable to trail game. Like causes produce like effects.

Some wild tribes are found that still posses the peculiar powers of wild birds and beasts. The polluted air of civilization has not reached them yet and their centers of Cosmic Intelligence are not dormant and rendered practically useless by the destructive action of polluted air, in which civilized man lives and labors from birth to death.

Uncanny Powers Of Indians

According to innumerable observers and historians, as well as Indian tradition, when the Spaniards arrived to take over South America, they found that the Indian races had an uncanny and supernatural ability for conveying and receiving accurate information over long distances. If we are to credit the apparently unvarnished accounts, it was as remarkable in its way as medal telegraphy or mental telepathy. An Indian could and often did know exactly how many men or horses were approaching long before they could be seen or heard. He could tell where or in what direction a friend or foe was traveling, and he could perform many more equally mysterious feats.

Dr. Juan Durand, who devoted many years to a study of Indian history, traditions and life, personally witnessed such feats. One night, while at an Indian hut at Raco, the Indian owner placed his ear to the floor and told Dr. Durand the exact number of men in a platoon of soldiers who were passing at a distance of more than three kilometers from the spot. Another Indian at Panao, without rising from his couch, stated the number of men on foot and the number of mounted men traveling on a distant road, and even told the order in which they moved and the direction in which they were going.

In 1896, while between Cayumba and Monson, Durand's Indian carriers deserted. Other Indians, without faltering or hesitating, gave the exact route the deserters had taken and followed them for eight days across deserts, mountains and rivers where there was no sign of a trail or spoor, often cutting across county, and found the deserters exactly where they had foretold.

According to historians and to Dr. Durand, the Peruvian tribes were able to receive such information of distant events by their ability to "read" the barking and howling of their dogs, and that this knowledge of the dogs' language thus enabled them to receive information and full details of matters of which they would otherwise know nothing. In all probability this was merely an explanation to satisfy the curiosity of the white man.

An amazing demonstration of the miraculous powers once active in the body, but apparently dormant in civilized man, occurred in Czechoslovakia and was recently reported in the Magazine Digest.

It appears two young men discovered that after certain vigorous breathing exercises in good, fresh air, they could make themselves into human radio receiving sets. With nothing but a loud speaker, on which they put their hands, they could at will tune in any station within several hundred miles and bring in the music clearly through the loud speaker. They were investigated by reporters and professors, but no explanation could be offered except the breathing exercises appeared to be essential in conducting the feat. There are inexplicable mysteries in the air we breathe and in the various organs of the body, about which Modern science knows nothing. We have learned a little about these air mysteries by the invention of the radio-radar-television mechanism. Previous to these inventions, no one had believed in the air mysteries that we have discovered. We will recover more of these lost and miraculous powers of the body as we resurrect its dormant and deranged organs by living more in harmony with cosmic law.

Carrel says that man is not confined to his body, but diffuses through space. In telepathic phenomena, he instantaneously sends out a part of himself, a sort of emanation, which joins a far-away relative or friend. He thus expands to great distances. He may cross oceans and continents in a time too short to be estimated.

The hypnotist and his subject are sometimes observed to be linked together by an invisible bond, which seems to emanate from the subject. When the communication is established between the hypnotist and his subject the former can, by suggestion from a distance, command the latter to perform certain acts. At this moment, a telepathic relation is established between them. In such an instance, two distant individual, are in contact with each other, yet both appear to be confined within their respective anatomical limits. Thought seems to be transmitted, like electromagnetic waves, from one region of space to another. We do not know its velocity. Neither biologists, physicists nor astronomers have taken into account the existence of metaphysical phenomena. Telepathy is a primary a datum of observation. We know that Mind is not entirely described within the four dimensions of the physical continuum. It is situated simultaneously within the material Universe and elsewhere. It may insert itself into the cerebral cells and stretch outside space and time, like an alga, which fastens to a rock and lets its tendrils drift out into the mystery of the ocean.

We are totally ignorant of the realities that lie outside space and time. We may suppose that a telepathic communication is an encounter, beyond the four dimensions of our universe, between the immaterial parts of two minds. But it is more convenient to consider these phenomena as being produced by the expansion of the individual into space.

In rare instances in the polluted realm of civilization, it still occurs that strange intelligence is shown by a child of perhaps five or six years of age, and the child is regarded as a prodigy. Modern science is unable to offer any sensible explanation of this peculiar phenomenon. It may be a case where polluted air has not yet had time to dull and dormantize the spiritual centers in the child's head, and it is able to contact and receive certain phases of Higher Intelligence direct from the Cosmic Source, as the Ancient Masters did ages ago.

In a few years, polluted air has done its destructive work, and the child who was once a prodigy, sinks in the realm of intelligence to the level of the social pattern of the masses. Thus, do we become what our environment makes us, while we in turn make the world in which we live.

Lesson No. 23
Spiritual Powers

"We possess no technique capable of penetrating the mysteries of the brain....Our intelligence can no more realize the immensity of the brain than the extent of the sidereal universe..... The cerebral substance contains more than twelve thousand millions of cells" (Carrel, in *Man The Unknown*. Pp. 9, 95).

That overt admission of the greatest medico-scientist since Darwin and Huxley clearly exposes the falsity of medical claims about solving the secrets of the chemical, physiological, psychological and biological operations of the body, so fearfully and wonderfully made (Ps. 189:14).

In a lecture delivered at Dornach, Switzerland, April 1st, 1922, Rudolf Steiner said: "In his head, in the wonderful convolutions of his brain, man is the image of the entire cosmos. In the body of the mother the human being is formed as an image and likeness of the Universe. Man is first brain, the image of the cosmos. We can study the cosmos by studying the human embryo in its early stages."

When Cosmic Radiation starts the formation of a new person, the process begins with the brain, and next with the nerves. A 26-day-old fetus consists almost entirely of brain substance. The body then looks like an elongated brain. The head of a normal, new-born baby is more fully developed than is any other part of the body, and develops less after birth than any other part. This fact indicates the relative importance of the brain. The brain, spinal cord and nerves are by far the most important parts of the body. These organs are found to be normal in persons who are said to have died of starvation. They are sustained by the oxygen and nitrogen gases in the air and the oxygen and hydrogen gases in the vapor in the air man inhales (Lesson 19).

It is computed that civilized man rarely develops more than ten per cent of his potential mental capacity because most of his thinking and brain work are done for him in order to bind him into the social pattern. Every part of the body and every organ and gland are under the direction and control of the brain, through the agency of the nerve system. Without the five physical sense organs and the brain and nerves, man could have no knowledge of the physical world, or of anything in it. He would not be conscious of his own existence. He could not be aroused from his slumber any more than a tree can become conscious of the animal plane.

Ancient Science Of Man

Modern science knows not that the correct interpretation of the first six chapters of Genesis and the Book of Revelation of the Bible, and the Ancient Science of Man, are concealed in his own body, and there preserved forever in the rudimentary organs, the embryological and homologus structures remaining dormant in his organism. People have been so completely blinded by the theory of Evolution, that they do not try to understand these strange signs of the Creative Principle, or to discover the hidden meaning. The rudimentary structures in man's body are not dead. They are only dormant, and are capable of rebirth, resurrection, regeneration and rehabilitation. When "born again" they will present that Physical Perfection in Man known to the Masters more than a million years ago.

The Spirit of the Universe is in the air man breathes. Deposited in his blood and nerves, with each inhalation, is the power that animates his organism and produces that intelligence which directs not only the mysterious phenomenon of Life itself, and the conduct of birds and beasts, but the course of the planets through the misty reaches of space.

Spiritual Intelligence

Spiritual Intelligence is a phenomenon so far beyond the comprehension of modern man that the very term means nothing to him. One author says that "Radar is the new scientific name applied to a most ancient occult practice." Birds live in the higher, purer currents of air. Their spiritual intelligence organs have not been crippled by the poison air of civilization, and much of their conduct is inexplicable. They fly in large groups, go in the same direction, perform in various ways, and wheel in the air as if controlled by one Great Mind. That is an example of the work of the Cosmic Mind in cases of animals whose spiritual chambers have not been crippled by poisoned air. The same delicate powers of electro-magnetism enable the birds to fly on a curve. These curves are derived from aerial magnetism, of which, so far, man is able to use only the north pointing vibratory rays.

It would be preposterous to suggest that man can make what God cannot. Man's skull contains the ruined remains of the most perfect radio-radar-television mechanism that one can imagine, with five tubes termed Sinuses, all of them dual purpose, with built-in radionet antenna, automatic power rectifier and automatic control. This mechanism in man's skull is the original pattern that has been imitated by the work of art. The imitation cannot begin to compare with the original. It is only an artificial replica of what man once had and lost, — his miraculous unused powers.

Man's brain and nerves are the physical mechanism that releases him from his physical tomb of silence and darkness, and gives him all the knowledge he has of his physical being and physical environment. Before his perfect radio-radar-television mechanism was ruined by poisonous gases, it released him from his physical senses and physical environment, nullified the illusion of space and time, and revealed his dual personality, to the effect that he is temporal in the physical and eternal in the spiritual. Then the Spiritual Light of the Cosmos illuminated the Field of Infinitude in man's physical consciousness, and his Spiritual Consciousness, becoming active, made him omniscient for that period. The past and the future, space and time, vanished and became for him the Eternal Present.

In the realm of Spiritual Intelligence, the first new psychic sensation is that of a strange duality in oneself. As this change comes, man finds himself in a world entirely new and unknown to him. It has nothing in common with the physical world. It has no sides or limits; all is visible at once at every point. Everything is unified, linked together. Everything is explained by something else, which in its turn explains another thing. To describe the first impressions or sensations, it is necessary to describe all at once. Should one attempt to describe the realm of Spiritual Intelligence, one has no worth for that purpose. Language that describes the physical world cannot describe the Spiritual. That is the reason why one who has had mystical experiences uses, for expressing them, those forms of images and words of the physical world. But these describe the physical world and not the Spiritual. Therefore, one who returns from the realm of Spiritual Intelligence, the mystical states of consciousness, cannot describe one's experiences because it cannot be done in the language of the physical world, and one knows no other.

Prof. Hotema   *Man's Higher Consciousness*   *Lesson No.23*

Man's Intelligence

Man's body is composed of trillions of cells. Each cell is composed of millions of atoms, each of which is a miniature solar system, with "planets" in the form of electrons whirling at tremendous speed round a common center of attraction. The cells of man's body are intelligentized by Cosmic Consciousness, and animatized by Cosmic Force.

Man's intelligence comes through his cells, direct from the Cosmic Source. Cosmic Intelligence is limited in man due to his limited capacity to receive and express it. This capacity arranges men into many classes; and those of each class express intelligence according to the condition of the body. The more perfect the body, the greater the intelligence it will express. Man is a creature of vibratory impressions received from Cosmic Rays. This makes modern man's capacity of consciousness very small in his present degenerate state. He is dependent upon his five degenerate physical senses to contact the radiations of the Cosmos and these senses are more or less deficient, while his Five Spiritual Senses have been dormantized and rendered useless by the poisoned air of his environment

Intelligence Of Animals

Naturalists tell us that they are perhaps five hundred other senses used by bugs, birds and beasts. Poisoned air has not damaged their sense organs. Ants, bees and caterpillars navigate by the sun or the moon. Their eyes can detect sunrays even through clouds. Gymnarchus Niloticus, a fresh water fish, sends out electric impulses at a rate of several hundred a second, which create an electric field – something that degenerate man has difficulty in detecting. The fish feels things at long distances in this manner. Birds of prey that see miles away, do the trick neatly with devices in their eyes that enable them to keep their sight fixed once they have seen something they want. Man can do this in reverse by watching an airplane vanish. He can see it much farther if he keeps his eyes fixed on it. Flashes of light by fireflies are code signals by which the males attract the females. Bats make super-sounds to guide them by echoes which few but they can hear. More surprising is the fact that some insects, which bats eat, have the ability to detect the super-sounds and thus escape the bats. Water beetles that skate on ponds move fast but never collide. Their sensitive legs feel the force of the invisible waves caused by the other skating bugs and that feeling tells them the direction to go to avoid a collision. The dragonfly's neck is its compass. Its head is large, and any object that changes its course bends its neck. Then receptors in the neck send vibrations to put the bug back on its correct course. Locusts have sensitive spots on their heads that detect any change in the direction of flight.

Following mysterious highways in the sky, migratory birds travel north and south annually. From nesting grounds in the far north, to the south they go for the winter. Birds have done this in North America since the Ice Age, yet science is still uncertain as to how they follow their precise schedules and parts of flight, returning year after year to the same places in the north and the south. Many birds fly tremendous distances, sometimes nonstop, over thousands of miles of Open Ocean, returning in the spring by entirely different routes. No one yet knows how they navigate. Small Asiatic birds, migrating between Siberia and India cross the 20,000-foot peaks of the Himalaya Mountains. The Pacific Golden Plover flies each fall 2,400 miles across an islandless course from Alaska to Hawaii finding its destination unendingly.

Long-distance champion of the bird world is the Artic Tern. Nesting as far north as there is land on the islands rimming the Arctic Sea, these birds fly in early September across the ocean to Europe, thence down the west coast of Africa, and eventually to the fringe; of the Antarctic Ocean, the south polar region. Returning in the spring via South America, the globetrotting Tern covers a distance of some 22,000 miles in one year.

In the field of television, Deslandres said that the homing-sense of birds appears to rise as to the effect of a mysterious electric perception. He wrote: "Birds can home over territory that offers no visible landmarks. I have seen a pigeon released from a balloon at a height of 5,000 feet. The bird was carried in a closed box. As soon as released, it rapidly described two circles round the balloon and then, without hesitation, darted off in the direction of its dovecot 250 miles away."

The press of May 9th, 1952, told of a cat that travelled 450 miles to get home. The item says that A. S. Snyman took his two year old cat from Groenfontein, South Africa, to Brandfort, in the Orange Free State, and left it there, returning home by car. The cat, very thin and about exhausted, showed up twenty-six days later at Snyman's farm in the Cape Province, about 450 miles from Brandfort.

In our course of study titled "IMMORTALISM," we go more fully into the subject of the strange powers possessed by the lower animals and formerly by man, but lost by him because of degeneration. In instances of degeneration the higher powers always fade out first.

Man A Miniature Universe

Paracelsus said: "Man, as microcosm, is formed of the same elements as the Universe, as Macrocosm."

In the Sankhya doctrine concerning the twenty-five elements of Being, we are told that man's five physical senses are only the exteriorized products of the five corresponding latent specializations of the primary ego-forming Conscious Essence or soul Substance — the Department of Eternal knowledge. So the Ancient Masters taught that as man is Microcosm, a miniature universe, all things contained in the Macrocosm are also contained in the Microcosm in character if not in degree.

The special sense organs in the perfect bodies of the Ancient Masters were normal and in sympathetic vibration with Cosmic Radiation, hence they could travel in a direct like as birds now do, toward a distant goal that would be invisible and unknown to Modern man. They were able to detect vibrations that our dulled, dormant, degenerated Spiritual Chambers cannot perceive.

Man Is Dead As He Lives

Consider man in a faint, or unconscious from drugs or anesthetic or injury. His body otherwise functions with normal activity to maintain physical life. Nothing is absent but his physical consciousness. The subconscious power, the inner, spiritual man, is intact, uninjured, unchanged, and active. It is only the physical aspect of the conscious mind that is inactive, functionless, as a result of which the open eyes can see nothing, the ears can hear nothing, and the physical powers of smelling, tasting and feeling are absent. With his five physical faculties inactive, closed and shut off from receiving any vibratory impressions conveying intelligence of the physical world, and also being unable to send forth any messages, the conscious mind of physical man is closed and dead to all physical existence. *Such man is literally dead as he lives, so far as his conscious contact with the physical world is concerned.*

Man Lives In The Spiritual World

Man is never more spiritually dead than when physically alive. His physical organs of higher function, which contact the Spiritual World, are dormant. Physcal man in a state of physical unconsciousness knows nothing so far as the physical world is concerned. Were he actually dead, he could not know less of his earthly being. Yet his body otherwise functions physically as though nothing had happened. He lives physically, but the physical aspect of his Mind is blank. While in that state, he could enter the Spiritual World and return from it, but know it not. In fact, he could be living all the time in the Spiritual World, and be unaware of it.

In the clever scheme to darken the Mind, the despots directed the scribe to write: *"Except a man be born of water and of the spirit, he cannot enter into the kingdom of God". (Jn. 3:5)* Then they neglected to delete the statement in the Luke that proves above statement to be false. It is written-*"The kingdom of God Is within you" (Lu. 17:21).*

The spiritual darkness of today is the product of a plan invented at the First Council of Nicea in 325 A.D. The darkness was produced, as then and there decreed, by destruction of the Ancient Wisdom and the invention of a new theology that enthroned the priesthood and enslaved the masses. That Ancient Wisdom taught the dual personality and the natural immortality of man by revealing the fact that he is temporal in the physical but eternal in the spiritual. Nature, internally and externally, is filled with light-spiritual through and through. So is man, her highest product. But his spiritual light is within, not without. It is the scheme to lead man astray by teaching him to look here and there for the kingdom of God, when he should look within (Lu. 17:21). Go into thy closet not to the cathedral, and examine the inner sanctuary of the throne of God, within the body, and there find the goal of all human desires (Mat 6:6).

The inner vision of the Masters was fully opened. They saw though Nature as through clear glass. They saw more through their mind's eye than through their physical. To them all Nature stood revealed to her inmost depths, wherein they saw One Essence, One Spirit, of which all things within and without are but various phrases of its manifestation. In it the Masters saw themselves as parts of the same phases, living moving and having their being sustained by One Spirit which is both Life and Light. The Masters knew they must live a perfectly natural life, in absolute harmony with the spiritual and physical laws of the Cosmos, and were liable neither to death nor illness as a result. They lived as part of Nature breathing in unison with the breath of sky and air, birds and beasts, trees and grass, their souls in tune with the Great Soul of Infinity Itself.

The best spiritual, intellectual and moral men of today cannot be compared with the Masters whom we, in our dense ignorance of facts of the remote past, like to call savages and barbarians to make them smaller and ourselves greater. This is all part of the plan of the despots to darken the mind and control the man.

Parthenogenesis (Virgin Birth)

In his book, "Sree Krishna," Bharati states that the Golden Age lasted nearly three million years, and "was the most spiritual age of man" (P. 65). Men were then physically immortal, with a Life-span of a hundred thousand years, and they died at will by sinking into a deep sleep, leaving the physical body behind and returning to their Spiritual Home.

"Men and women had no need for sex life in the Golden Age," he says (P. 68). The Law of Parthenogenesis ruled, and sexual generation was unknown — as so well described by Dr. George R. Clements in his masterful work "Science of Regeneration." "During the Golden Age and the greater portion of site Silver Age," says Bharati "all men and women were what the Christians call virgin-born." (P. 136).

When the Romanite missionaries carried to India in the 5th century the virgin birth story rated in the Matthew and the Luke, but entirely omitted from the Mark and the John, it did not move the Hindu Masters. The Romanites were utterly ignorant of the fact that they had invaded the land from whence they had received the virgin birth tradition. The gospel story was based on a report taken from India to Asia Minor by Apollonius, who visited India between 36 and 38 A.D. and again between 45 and 50 A.D., and there received the account of the virgin birth. He brought back the Hindu religion and founded a communistic sect of Nazarites at the village called Nazarita. According to ancient history, the Hindu god Chrishna was said to have been born in 3333 B.C., of the virgin Devaki. Concerning this Sir William Jones wrote: "In the Sanscrit dictionary, compiled more than 2000 years ago, we have the whole history of the incarnate deity, born of a virgin, and miraculously escaping in infancy from the reigning tyrant of the country" (Asiatic Researches, Vol. I, P. 273) Regarding the virgin birth, Bharti said: "The fuss that is made about this immaculate conception (by the Christians) succeeds only to excite a smile of pity in the Shestra enlightened Hindu — a smile of pity for the ignorance of the facts in the past history of humanity, of which they seem to know so little and care less to know more". (P. 136)

The Kingdom Of God

Ancient Science taught that man lives here, now and always in the kingdom of God; as that kingdom includes all, both the physical and spiritual worlds. There is One World with dual aspects: The Spiritual and the Physical. Man actually lives here and now in the Spiritual World, but knows it not because his mind is darkened by the teachings of established institutions, and because he is unable to contact the Spiritual World due to the degenerate state of his Spiritual Radio — the Air Chambers in the skull. Due to degeneration, the Five Spiritual Faculties of civilized man, of which his five physical senses are only the exteriorized products, are dormant, closed, inactive, shut off from receiving the vibratory currents conveying to him the cosmic intelligence of the Spiritual World.

In wild birds and beasts, under the control of their vital powers, the organic currents of their highs vital activities tend towards correspondence, towards contact with the cosmic currents and forces. Such animals have instinctive knowledge of everything connected with the preservation of their being, and such knowledge was possessed by primitive man in even a higher degree. This explains many phenomena as to the conduct of these animals which scientists should be able to understand, but do not. Grasshoppers have weak sight, but their intelligence organs enable them to find distant regions rich in vegetation. Fishes find suitable spots to lay their eggs. A cat or dog tied in a bag and taken a long distance by train will return to the home some weeks later. A turtle found at Ascension and taken to England made its way to the English Channel and was found again two years later at Ascension.

The Tampa (Florida) Tribune of October 16th, 1951, states that in 1949 Mr. and Mrs. C. D. Smith moved to California after selling their home in St. Petersburg, Florida. They gave their cat "Tom" to the buyer, Robert Hanson. Two weeks later, Hanson wrote them that "Tom" had run away; and that was the last the Smiths heard of the cat until two years and six weeks later, when Mrs. Smith heard a cat meowing in their California yard and asked her husband to go out and chase it away. He went out to do that, but instead of running away the cat leaped into his arms and began to purr. "Hey, Betty, Smith shouted to his wife, this is old Tom." Old Tom was skinny and worn from his 3000-mile hike, but happy to find his owners. Primitive people, in their unpolluted jungles, have the same extraordinary capacity for orienting themselves. It takes the polluted air and environment of civilization to destroy the spiritual powers of man.

For weeks before the terrible upheaval in Martinique in 1902, the population of St Pierre, the capital city, lived in a state of panic. The volcano called Pelle showed signs of eruption. Scientists and geologists made careful investigations and assured the people there was no danger. The population remained quiet, but the animals did not; first of all the amphibians, next the mammals, and last of all the birds left the area. A day later the terrific eruption occurred, and the whole population snuffed out by carbon dioxide. Their natural powers warned the animals of the approaching danger, while degenerate Homo Sapiens was last in spite of his science and his scientific instruments.

The same thing has happened in many other disasters and will continue to happen. Wild animals leave the dangerous areas and are saved, while man, not being warned, remain behind and perishes. Man's lost capacity to contact natural forces and cosmic currents is due to degeneration resulting from an artificial mode of living under artificial conditions in an artificial environment. His degenerate body is incapable of emitting or receiving the higher currents. He is incapable of receiving the electromagnetic vibrations that give warning of cosmic dangers because the unnatural life he has led for thousands of years has degenerated the appropriate organs. In view of his higher development, he should get many times better results than the animals do and the Ancient Masters did. When man losses his spiritual consciousness, due to the degeneration of his Spiritual Chambers, he is spiritually dead while physically alive, so far as his conscious contact with the Spiritual World is concerned. The Cosmic Essence that animatizes man's body and intellectualizes his mind, contacts the earth in that subtle, invisible compound termed Air, a substance concerning which modern science knows little, as we saw in Lesson 21, yet so potent and paramount that "the Essence of the Universe" is in the Infinite Air in eternal movement which contains ALL in itself (Anaximenes). As this potent substance flows into man's lungs through his nostrils, a certain portion enters and activates the "inner shrines" of the Temple of God, situated in the Golden Bowl (Skull), and termed Sinuses. In civilization this potent substance, without which man would die in a few minutes, is poisoned beyond description by his works and inventions. Polluted air fills the cities and homes, shops and hospitals, and deterioritizes and dormantizes the nerves extending to the Brain from the Five Sinuses, the "Inner Shrines."

As the function of smelling weakens and fails because polluted air has deterioritized and dormantized the nerves extending to the brain from the nose, so the spiritual function of the sinuses weakens and fails for the same reason, and man thus becomes dead as he lives, so far as the Spiritual World is concerned.

Feeble Minds

Dr. Charles H. Mayo said: "Every second hospital bed in the United States is for the mentally afflicted." Professor Eli G. Jones. MD., a prominent American physician and educator with fifty years experience, said that every fourth physician was a drug addict. Civilized man has not only lost practically all his spiritual powers of contact with the Spiritual World, but he is rapidly losing his physical powers of contact with the physical world.

Carrel declared that in New York State one person out of every twenty-two must be placed in an asylum at some time or another. In the whole of the USA the hospitals, care for almost eight times more feeble-minded or lunatics than consumptives. He continued: "In the whole country (U.S.A.), besides the insane, there are 500,000 feeble-minded. In addition, surveys made under the auspices of the National Committee for Mental Hygiene have revealed that at least 400,000 children are so unintelligent that they cannot profitably follow the courses of the public schools. In fact, the individuals who are mentally deranged are far more numerous. It is estimated that several hundred thousand persons, not mentioned in any statistics, are affected with psychoneuroses. These figures show how great is the fragility of the consciousness of civilized men" (*Man The Unknown*, P. 155).

According to the press of November 28th, 1947, psychiatrists estimate that one in sixteen persons in the U.S.A. is mentally weak. The report says that Dr. Vladimir Eliasberg stated that there are 800,000 insane persons in various institutions, and eight million more are wandering through the cities, because their families and friends consider them harmless eccentrics and let it go at that. Millions more are in the beginning stages of paresis. Man is dead to the physical world when his physical senses fail, although still alive. He is dead to the Spiritual World when his spiritual senses fail, yet he is still alive. Civilized man has lost all contact with the Spiritual World, and is rapidly losing contact with the physical world. Much importance is given to physical food while no attention is given to spiritual food. Physical food is what we eat, while spiritual food is that which we breathe. While only a child, the poisoned air of civilization deteriorates man's body, dulls his physical senses, and dormantizes his spiritual powers. He makes his world and his world makes him. Perfect Correspondence must prevail. That is cosmic law.

Ancient Science Vs. Modern Nonsense

Modern science holds that man is only a physical body that functions as the result of 'a series of chemical changes'. When body function stops, man's extinction comes. Ancient Science taught that the body is a material instrument, and as such it is: 1. Constructed by Cosmic Processes 2. Animated by Cosmic Power 3. Directed by Cosmic Intelligence.

According to ancient philosophy, the Cosmic Breath of Life enters man's body through his nostril (Gen. 2.7). The channel provided for that purpose, and passes into the Spiritual Chambers of the head and on to the lungs. In the lungs it is absorbed by the blood and carried to every body cell, while Cosmic Intelligence, through the subconscious mind, directs the body's involuntary functions. That ancient doctrine is a fact of observation, while the theory of Modern science is the fancy of a weak mind. *If man were taught the natural rules of health, and lived accordingly, he could not be sick. The body functions always in the direction of health, otherwise the sick could not recover. The lawful function of the body must be obstructed before sickness appears, and the* symptoms of sickness rise from the body's effort to remove the offending obstruction. To suppress those symptoms is dangerous and may result in death. Those who have finished by making all others think with them have usually been those who began by daring to think for themselves." – Colton. "The old thoughts never die; immortal dreams outlive their dreamers and are ours for aye; no thought once formed and uttered ever can expire" – Mackay.

Lesson No. 24
Physical Purification

"*The blood moves through the arteries at the average rate of thirty feet a second. When you sit quietly, about five pints of blood pass through the heart each minute. That quantity is increased to thirty-five pints each minute if one runs uphill.*" — Dr. Arthur Vos in Health Messenger.

'Breath is life.' wrote Pundit Acharya. The Breath of Life is the Power of Animation (Gen 2.7). It fills the body with mysterious force that defies modern science. It moves the blood, makes the heart contract and expand, builds the body, and vitalizes and intelligentizes it. We can trace the course of the blood vessels, analyze the composition of the blood, show the action of the muscles, recognize the function of the cells, but no man can explain what Life is; whence it cometh, nor whither it goeth. We can follow the blood stream to the capillaries, and there we must stop. We can trace the course of the air to the air cells of the lungs, and there we must stop. We can determine the amount of oxygen absorbed by the blood and the amount of carbon dioxide excreted, and there we must stop. That is the limit of human knowledge in this field. Beyond those points modern science has been unable to go. The manner in which Life Force animates matter, in which cosmic rays and sunlight are condensed into blood and bone, in which breath and blood possess the sustaining power and productive properties — all this work occurring every second of time under our very nose, and yet they are cosmic secrets, and seem to be beyond human ken.

All mankind stands in dumb silence and humble worship before the mysteries of the Cosmos, and modern science is no less confused than the man in the street. That Power which moves matter, makes substance more than form, gives condensed gas vitality and intelligence, enables matter to see, hear, taste, smell, feel, think and reason — these Seven Mysteries of Life — that Power has not yet been seen, nor heard, nor touched, nor named. Yet science claims that Food is the Source and Power of the Vitality, Intelligence, and Strength of the body. Science has formulated that theory and now must support it. To repudiate it would leave science empty and void, — a system of speculation without a stone on which to set.

Function Of Breathing

Life is not individualized in the Infant until it inhales the first breath. Until that moment, the vitality pulsating in the Infant's body meets through the mother, but not from the mother. The moment the new-born baby inhales the first breath, its body becomes capable of that function which constitutes living-growth, replacement of cells, repairment of fractures and all other injuries, and recovery from the various ailments when not hampered by doctors and their poisonous remedies.

Every living thing must breathe the Air of the Universe or die. Trees breathe through their leaves. Thus, the leaves are the lungs of the tree. Insects breathe through tiny openings in their bodies. Frogs breathe partly through the skin. Fishes breathe by absorbing oxygen out of the water as it passes over their gills. Man breathes partly through the skin, but largely through the lungs.

The term Blood Poison may frighten you. But in the process of living man's blood is poisoned by a constant process body function. It is just as constantly purified. The purification process not only occurs in the lungs, but that is the only means provided by the Cosmic Creative Process to purify the blood. If the layman knew that, it would mean a big loss to those who live and thrive on human misery. It would stop the annual expenditure of millions of dollars for worthless blood tonics and blood purifiers. When the blood flows from the heart to the lungs to cast off its cargo of poison, and be purified by the air in the lungs, the blood is then and there further poisoned by the polluted air in the lungs. Read that again. The capillaries are a vast network of blood vessels that connect the arteries to the veins, and so small that they could not be seen with microscopes in use in 1816 when the celebrated Harvey discovered the circulation of the blood. He was task to describe how the blood passes from the arteries to the veins. Were the heart a pump as claimed, it would have to force the blood from the feet up to the heart against the pull of gravity. The pressure of this large volume of blood would fall on the tiny capillaries, with walls thinner than a soap-bubble, and the entire capillary system below the heart would be ruptured and bursted in an instant by this back pressure. The capillaries in the lungs are the last tubes through which the blood flows to get the air gases it must have, or fail in its function. Sixty to 80 times a minute it flows from the heart to the lungs, and it must always find air there waiting for it, or death ensues.

The lung capacity is so large that an average man inhales daily approximately 777,000 cubic inches of air, and in the same time 125 barrels of blood pass through the lungs for purification. In the lungs there are millions of capillaries. They twine among the tiny air tubes and air cells as a vine twines among the branches and leaves of a tree. The walls of the little breathing organs of the tiny capillaries are much thinner than the walls of soap bubbles. The thinnest film imaginable separates the Blood and the Air in the lungs. It is here that the ultimate act of breathing occurs. It is here that the air and blood intermingle.

It is here that the Breath of Life passes into the blood. It is here that the poisons, the filth and impurities of the body are brought by the blood and cast off, and a new load of oxygen, Nitrogen, Hydrogen and the Essence of Sunlight is absorbed by the blood and conveyed to all parts of the body to furnish the trillions of cells with the normal stimulation to activate their various functions. The slightest interference with this vital process is fatal. The lips quickly turn bluish-purple when respiration is obstructed, due to the rapid collection of carbon dioxide gas in the blood. In just a few seconds the blood would turn almost black in color if respiration is obstructed or halted.

Shower Of Red Mist

The mass of blood vessels in the lungs are distributed everywhere in the minute spaces between the millions of air vesicles, and envelop their walls with a vascular network. The blood flows through the lungs in thousands of minute streams, almost in contact with the air contained in the vesicles. In fact, it is as though the River of Living Water were sprinkled through the breath of Life in an exceedingly fine shower of Red Mist, so that every tiny particle of blood and every atom of the Breath of Life in the lungs are brought together in the closest proximity.

The entire blood supply of the body passes through the lungs for purification many times each hour from birth till death. As the blood enters the lungs it is a dark blue or purple color approaching to black. This is the venous blood and it is loaded with all the filth and poison collected from the cells, tissues, glands and organs of the body. As this blood enters the lungs, it is stream of poison in every sense of the word.

This is the blood that flows back to the heart through the great veins from all parts of the body, to flow on to the lungs for purification.

A marvelous change occurs in the color of the blood as the purging process occurs in the lungs. At this instance, the air gases in the lungs are absorbed by the blood over the whole internal surface of the lungs, the dark, poisonous stream is changed in color, as though by magic, to a brilliant scarlet. This is Blood Purification, and this is the only way in which the River of Living Water can be purified.

Blood Poison

The Vital Stream that turns the Wheels of Life is not only the health-producing and life-sustaining agent of the body, but it is also the destroying power. It could not be otherwise without reversing law and order in the body's vital economy. From the millions of air cells in the lungs, the gases in the air we inhale pass into the blood, and are collected by the red blood corpuscles. They are about 1/3200 of an inch in diameter, and the blood contains 25 to 30 trillions of them. Their total combined surface would cover an area approximately 200 feet square. When we inhale polluted air, the red corpuscles recoil from it in the lungs because it is dangerous, and then trouble begins.

The red corpuscles have a double concave surface, and a smooth outline at their edges. The absorption of poisonous gases and fumes into the blood through the lungs causes rapid changes in these corpuscles. They lose their roundness, becoming oval and irregular; and instead of having natural attraction for one another and running together as they do in good health, they lie loosely scattered before the eye, and indicate to the learned observer as clearly as though they spoke to him, that the one from whose blood they were taken, is physically depressed and deplorably deficient both in mental and muscular tone.

The tiny capillaries in the lungs are just large enough to allow the red corpuscles to flow through them in single file. The only element that separates the corpuscles from the air in the air-cells of the lungs, is a thin membrane about sixteen-one hundred thousandths of an inch thick. If we apply heat to the skin, the blood vessels in it expand and become red. If we inhale air too warm, the blood vessels in the lungs expand, and the corpuscles cannot pick up oxygen so easily. That is one reason why very warm air is suffocating. Again, if the air is very poisonous, the corpuscles recoil from it, which also produces a suffocating sensation. The symptoms of suffocation are not always the result of the conditions mentioned. They may appear because thickened or carbon-coated walls of the air-cells prevent free passage of oxygen into the blood.

Cold air is bracing because it does not expand the blood vessels in the lungs and the corpuscles can readily absorb the oxygen. Also, cold air contains more oxygen than warm air. The air contains 25% more oxygen at zero temperature than at 100 degrees above zero. Symptoms of suffocation appear in chronic and semi-chronic lung ailments, as asthma and tuberculosis. Labored breathing is one of the chief conditions of old age. It is not the cause of old age. A man of 50 should breathe as easily as when he was 20. Why he does not is what we are going to learn.

Breath Culture

The Broom publication contained in a certain issue, issued the following remarks under "Breath Culture": "When you deal with Breath you are dealing with Creation — with the power that builds and destroys. You can create both ways: Heaven or hell. I have studied and practiced Breath Culture, consciously and designedly, for over forty-five years. Take a locomotive, a steam engine; it pulls and breathes, and with every Exhalation the steam hits the cylinders (piston heads), which make the wheels go round. Take the automobile — with every exhalation, the pistons are hit by force, which makes the wheels go round.

"The dynamics of the body are in the lungs. No breath, no motion, no life, no thought. The steam engine, with all its puffs and snorting, does not think. It moves, it pulls loads, it travels fast, but it stays within the limits of the law set by the designer and travels on fixed rails. Man is a different engine. He thinks, but he can think only as the rails or ruts will let him think. (Note: His mind and education are controlled and he thinks according to a pattern prepared for him. — Klamonti).

"As long as you bear in mind that you cannot think at all, while in this human frame, without grey matter, vibrations and chemical changes in the grey matter, you will not think of thought metaphysically, but intelligently, realistically. Spirit comes from Latin 'spirare,' to breathe. Thus spirit is a very material process too, my friend, not merely metaphysical." (Note: There is a vast difference between Spirit and

Breathing. Breathing is a mechanical process while Spirit is the substance inhaled: Prof. Hilton Hotema).

It is not an empty allegory to assert that the gaseous elements termed Air, God's spiritual substance flowing into man's lungs, is the steam in the boiler of the human locomotive that makes it move and supplies the power exhibited by the body. The average adult inhales 480 cubic inches of air per minute while at rest; five times as much if he walks four miles an hour, and seven times as much if he walks six miles an hour. It is not food but air that supplies the body with power.

Vital Function

God is the Master Economist and makes nothing in vain. There is a scientific reason why He made man's lungs so large. Air is so important to living that the size of all the body's organs sinks into utter insignificance when compared to the lung. While man is careful about what he eats, he pays no attention to the kind of air he breathes unless it is so foul as to be nauseating.

In a normal pair of lungs there are approximately a billion tiny air cells. If they were all spread out in a flat surface, they would cover a space about 40 by 50 feet. This is the breathing surface that directly contacts the air in the lungs. This is man's vital capacity. All the air in the lungs is not changed at each breath. Normally, we inhale about 30 cubic inches of air, or about 500 cubic centimeters each time we breathe. This is called the (1) tidal air. It comes and goes without special effort. If we take a deep breath, we will inhale 100 cubic inches more. This extra cubic intake is called the (2) complemental air.

Vital powers. — Suppose sudden danger arises, as in the case of an angry bull, and we need extra energy to flee. We are told energy comes from food combustion. But our stomach is empty, and we cannot depend on food for extra energy in the emergency. And if we could not at once inhale more than 30 cubic inches of air, the nerves could not supply the vital power necessary to make our legs move at top speed. In such cases, extra air capacity is provided by greater lung expansion and faster breathing. At the end of our run we find ourselves breathing hard and fast. This extra air is termed the (3) reserve or supplemental air, and amounts to approximately 1200 to 1500 cubic centimeters. Besides the tidal, complemental and supplemental air, there is a certain amount of air that always remains in the lungs. No matter what we do, we cannot force all the air out of the air cells of the lungs. If we could and did, we would drop dead.

The Residual Air

About 100 to 200 cubic inches of air constantly remains in the lungs after the most violent expiratory effort. The amount depends in great measure on the absolute size of the thorax, but may be estimated at about 1,000 to 12,000 cubic centimeters. This is termed the (4) residual air, without which the function of the body cells would fall below the life level, and that would be somatic death. The Residual Air in the lungs is all that stands between life and death. It is the vigilant guard that protects man from the dangers of – 1. Very cold air that would kill quickly if it could enter the terminal air sacs without first being warmed by the residual air, and – 2. Very dirty air which, if sucked directly into the terminal air sacs, would coat their walls so thickly with filth that sufficient oxygen to preserve life could not pass into the blood, and one would die quickly of suffocation.

If the air of zero weather and colder could be sucked right down into the terminal air sacs of the lungs, the sacs would freeze immediately, and man would drop dead as the penalty for living in such a hostile environment. The cold air must meet and mix with the warm residual air in the lungs before it can enter the terminal air sacs. That is what stands between life and death in the case of those who live in cold regions.

The protection against dirty air prevents man from dropping dead in the filthy air of civilization. It cannot prevent him from dying by degrees from the cumulative effect of that filthy air. When we cough out dirty mucus, that dirt is in the air we inhale. The residual air prevents the dirt from entering the terminal air sacs, and we cough the dirt out as it accumulates in the lungs, provided it is free and does not stick to the walls of the lungs.

Gases And Acids

The Residual Air in the lungs warms the cold air as it enters, and to a certain degree obstructs dust and soot from entering the terminal air sacs. But it cannot stop the poisonous gases and acids in the air of civilization from passing directly into the terminal air sacs, and through their walls into the blood. In this category come the deadly gases of modern warfare, which were used in World War I to kill soldiers and others who inhaled it. Those not killed in battle by the gases did not live long, and none ever recovered health who had been seriously "gassed."

The deadly gases swept over Europe, causing an epidemic of lung ailments unprecedented in human history. The air in those days was so heavily charged with the gases and acids, that they were carried across the Atlantic by the winds, which may have resulted in the influenza and pneumonia epidemic in this country in 1918-1919, when thousands died of lung ailments.

For some years after the war, thousands living in the war zone of Europe suffered from lung ailments and many died. Could the gases and acids rising as vapor from the ground of the battle fields have been the cause? A report of the registrar-general of England in that 1918-1919 epidemic, gave a total of 112,239 deaths, or a mortality rate of 3,129 per million populations — the highest ever recorded.

In the days of the terrible cholera epidemic of 1849, which coincided with the wholesale vaccination of the population, the mortality rate per million was only 3, 033 (Journal of the A.M.A. September 11th, 1920, page 755). An article in the Scientific American for September 20th, 1919, stated that up to March 10th, 1919 — 56,991 U. S. soldiers had died of disease as against 48, 909 killed in battle. Of those that died of disease, 47,500 were charged to lung ailments. In other words, lung ailments killed as many soldiers under the best medical care this country could furnish, as weapons of the enemy killed in battle.

Thomas Frances, Jr., M. D., in writing of the 1918-1919 influenza epidemic, said, "In a period a few months, 20 million people perished, 548,000 in this country alone" (Journal of the A.M.A. V. 122, P. 4, May 1st, 1943). That made a death-rate from lung ailments of 5,211 per million population, an all time high that greatly exceeds the mortality rate of the terrible scourges of the dark ages. The horrible experience in the use of poisonous gas in World War I shocked the world, for so little is known about the agencies of death that float in the air, and caused the nations to agree not to use gas in future wars.

The Skin

The rapid manner in which the end product of cell function pollutes the body and the high importance of physical purification is well shown by the function the skin plays in the process. The skin is a porous covering of the body that is connected with a vast network of nerves, arteries and veins, and it contains billions of tiny openings called pores. The skin is an organ of elimination of poisonous substance resulting from cell function, and of assimilation of vital elements from the atmosphere, when it functions properly.

An historical event occurred that illustrates the importance of the skin in this work. During the inauguration of festivities of one of the Popes of Rome, a little girl was painted all over with gold paint so she would impersonate a cherub. Within twenty minutes she was dead. The cause: Automatic poisoning of the body by the poisonous products of cell function, consisting chiefly of carbonic acid gas, which could not pass off through the skin because the paint dosed the pores. This case shows how poisonous to the body is, the carbon dioxide is eliminated through the lungs and skin. It also shows that the lungs are greatly aided by the skin in the elimination of this deadly gas.

Exhalation

Protestor H. H. Sheldon of New York University, erected an apparatus in the Times Square theatrical district that drew in air at roof-level. In one week the apparatus cleaned 341,250,000 cubic feet of air, from which it removed 12 cubic feet of solid matter, composed of dust, soot and tar that weighed 37 pounds. The constant inhalation of such air results in a coating on the walls of the lungs, and in time one finds it hard to breathe. One puffs and pants from a little exertion, gasps for breath, and may have sensations of choking.

All the protection one has against polluted air is exhalation. The more vigorous the exhalation, the more poisoned air is cast out of the lungs. But this can accomplish nothing if one lives and labors in poisoned air. Animals know by instinct the value of exhalation. "The horse blows its nostrils, as does the dog, monkey or any other animal," writes Professor Godfrey Rodriguez in his "Key to Life." He said: "What animal has more strength for its size than a bull? When he blows his nostrils, it reacts like a fountain of force. The bull knows that the more he exhales, the stronger he gets. He will not take a chance with his nose alone. He is continually blowing, and the more he blows, the larger his chest grows, the smaller his waist-line, and the more poisonous waste be eliminates from his inner body through his lungs" (P. 9).

In the function of breathing, exhalation casts out the foul fumes and creates a vacuum in the lungs. That process is aided by and increased by coughing and sneezing. How to help Nature: Cough hard, from the bottom of your lungs, then hold your breath, and the vacuum thus produced in the lungs will exhort a pulling power that draws more poison from the blood into the lungs for elimination. Do this: Exhale to the utmost limit, blowing your breath out as long as possible. Increase the exhalation by hard coughing. Then hold your breath as long as you can. This creates a suction in the lungs that draws still more poison out of the blood.

Ernest T. Seaton, in his story of the Coyote, says that old trappers know it is true, that when a coyote eats poisoned bait, it is wiser than man, for it knows there is put one way it can overcome the poison, and that is by vigorous exhalation. It knows by instinct what you cannot even teach some people. If the poison does not get in its deadly work before the coyote can take a long run, and if it can run long enough and breathe fast enough, its lungs will eliminate the poison and save the animal's life. The poison has passed beyond the coyote's power to vomit it up. It cannot be eliminated through bowels or kidneys, and the coyote has no skin pores to sweat it out. So the lungs are the only channel of elimination. In most cases, the coyote beats the poison by exhaling it, just by eliminating it through the lungs. When the suffering body strives to cough out the poisonous air of civilization the usual practice is to take something to stop the cough. The body's functions are such a mystery to many, they do not know that coughing is a natural process of emergency elimination, the purpose of which is to expel anything that should not enter the lungs, whether it be a physical object or invisible gases and acids. A cough is a sudden expulsion of air from the lungs, and the velocity of the air of the human cough as it leaves the throat has been measured at more than 245 miles an hour. It may well be called a super-hurricane, and results from the tremendous pressure built up when the inhaled air becomes compressed in the lungs before released.

Coughing and sneezing are two emergency but natural processes of violent exhalation, of violent elimination, by which the body performs the important function of expelling poisonous gases and acids from the lungs by convulsive motions that send these gases rushing outward from the lungs with much force. Stupid is that person who attempts to hinder these beneficial processes. Remove the cause, polluted air, and the effect, coughing and sneezing, will subside.

Man would avoid much misery if he was taught that coughing and sneezing are beneficial, emergency, eliminative processes. Then one would know how to cooperate with these processes of purification and not obstruct them. On the contrary, he is taught to use poisonous remedies to stop a cough. While he continues to inhale the polluted air which the lungs are trying to drive out by coughing. As the remedy "cures" the cough, he remains in the polluted air. He inhales it; it passes into the blood and poisons the whole body. If influenza, pneumonia, and death then result, he knows not that it was the work of the poisonous remedies.

While in a certain room, you suddenly sneeze or cough without any apparent reason. As Nature's signals of danger are unknown to you, they go unheeded. Could you read these signals, you would know the cause and immediately seek better air. If you sneeze several times in succession, or continue to sneeze for some time, you are told it is "hay fever," contracted where there is no hay.

In the Pathometic Journal Dr. J. A. Little wrote: "I have not contacted any one teaching the importance of exhaling. The doctors have been following the old precepts of breathing that have been taught down through the ages, that is, to get more air into the lungs, force it in if necessary. The proper thing to do is to get the old, stale air out of the body by forced exhalations, to make room for fresh air. When we get the air out of the lungs, we need to make no effort to get fresh an in. It will rush in to fill the vacuum. We cannot prevent it from doing so, for the air outside of our bodies has a pressure of fifteen pounds to the square inch, and when we make space for it inside the body, the air will rush in to fill it. It is my contention, proven through many experiments, that the stale, foul air in the body has much to do in influencing our findings on the Kathoclast. Foul air lowers vitality; fresh air raises it. A clean body vibrates at a much higher potential than does a body that is filthy on the inside."

Atmospheric Pressure

Atmospheric pressure at sea-level is calculated at 14½ pounds per square inch. An average sized man supports with his body a pressure of 38,570 pounds, equal to a solid cube of lead four feet high. Under such pressure the air rushes into every vacuum, and man does not need to worry about deep breathing with such external power to push air into his lungs. He should worry more about exhalation. When he coughs and sneezes, he should do so with much vigor, to push the poisonous air out of his lungs against atmospheric pressure. When it remains in his lungs it passes into the blood and is carried to all parts of the body.

500 Felled By Gas

The press account stated that escaping chlorine gas felled more than 500 persons at a busy intersection in Brooklyn. Most of those overcome were quickly removed to hospitals, some in a serious condition. But no fatalities were reported. The gas, leaking from a tank being moved by truck to a Brooklyn pier, spread about two blocks in every direction. People on the streets began to cough, sneeze, vomit, stagger, then fall flat, creating a scene resembling war pictures. The gas, heavier than air, sank through ventilators into the subway, forcing people to flee to the street, where they also toppled over.

A man who was an eye witness, said he was walking along the street when people around him suddenly began falling like flies. Like some others who did not immediately feel the effect of the gas, he was shocked by the sight, and said the fallen victims looked like dead soldiers. Not knowing the cause of the trouble, this man tried to help those nearest him, when he suddenly grew sick. He began to cough and sneeze, his eyes to water; he began to choke, got dizzy, and fell unconscious. But if not for quick help, that would have been the end.

Women grew hysterical. All were coughing, sneezing, choking. They had pains in the chest, were dizzy and staggery like a man drunk. More than 20 hospital ambulances soon converged on the gas swept area, and provided inhalator treatment for the victims where they fell. A detachment of 100 gas-masked soldiers aided the rescue. Police, firemen and other, aided in the work. Some 300 victims were taken to hospitals.

Note the symptoms of coughing and sneezing — Nature's emergency processes of eliminating the fumes from the lungs. The usual method of treatment is to stop the cough and this is called "aiding nature." Stopping the cough, poisoning the sick is not "aiding nature." This Brooklyn incident further shows that the air of the cities is so saturated with poisonous gases and acids at all times, that just a little more added is all that is needed to send city dwellers to the grave. As it is, city air contains enough poisonous gases and acids to keep city dwellers coughing and sneezing much of the time, and in the sickbed a number of days each year. The ultimate result is early decay and early death.

Suffocation

Mrs. R. M. J. of New York recently wrote us as follows: "What causes me to feel like I'm suffocating just as soon as I am about to go to sleep? I have to jump out of bed, shake my head and press on my throat, with a feeling that I cannot swallow nor breathe. Then my heart beats wildly. Is it a nerve condition? My doctor does not understand it. Eating fruits and vegetables and fasting does not help. Please answer and explain if you can." These things are easy when one understands the underlying principle. The sensation of suffocation, gasping for breath, shortness of breath, labored breathing, mean three things: 1. Lung degeneration, 2. polluted air, and 3. insufficient oxygen. The only remedy on earth is God's pure, outside air.

Cold Facts

In January 1940, the press reported that at Amsterdam, Netherlands, ten children were cured of bad cases of whooping cough by an airplane flight of 90 minutes at an altitude of 10,000 feet, in the pure, ozonated air. Following that remarkable experience, hundreds of parents sent appeals to air line authorities to take their suffering children for flights, to cure them of whooping cough. These appeals were ignored.

The press of December 10th, 1938 reported Dr. H. Carlson as asserting that an airplane ride was foreseen as a cure for the common cold. and added: "Pilots, stewardesses and other persons who have much to do with airplanes, believe that a high flight would cure a cold. We made investigation among passengers, and found some 50 of them, who left Chicago with colds in various stages from the sniffles on, arrived in Newark, New Jersey airport with their cold entirely gone."

It was reported in the press of 1943 that some 26 persons, on a plane from New York to Los Angeles, had bad colds when the flight started, and all the colds were cured by the time Los Angeles was reached.

Whooping cough is a big mystery to medical art. The medical dictionary describes the symptoms and says it is "very contagious". Why is whooping cough confined largely to children? The better the lungs, the deeper the cough. In very early childhood lung degeneration is slight as a rule. So the cough motion, to eliminate polluted air, begins deep down in the very outer regions of the lungs. That gives the cough the peculiar, deep 'whoop' sound. The shallow cough is the cough of adulthood. By the time adulthood is reached, the outer-fringes of the lungs have decayed and lost their function because of the damage done by polluted air. Then one has the weak, shallow cough. The weaker the lungs the weaker the cough. Whooping cough in adults would indicate good lungs. That condition is seldom found in civilization, where poisoned air does its deadly work early in life.

Lesson No. 25
Breath Of Death

"Without the meeting of the air and the blood, the life of the Temple of God would end at once. Hence it has been so arranged by Infinite Intelligence that the air and the blood cannot fail to meet. When the River of life, dark with poisons, flows from the right ventricle of the heart through the pulmonary artery into the lungs, it always finds the air waiting in the tiny breathing rooms." F.M. Rossiter, B.S., M.D., Story of the Human Body, P. 124.

What kind of air does the blood find waiting for it in the tiny breathing rooms of the lungs? Man knows that he must breathe to live, but he thinks the KIND of air he inhales is not important. We saw in Lesson No. 21 how little is known about the Breath of Life (Gen. 2:7). Less is known, about the Breath of Death. It appears it is not known the function of Respiration is a dual process of life and Death.

Inspire, to live, carries Life into the Body; and Expire, to die, carries Death out of the body. Like all cosmic processes, this one produces results corresponding to the condition supplied. Under the same condition the same result is obtained. Under a change of conditions, it is evident there must be a corresponding change in results.

Man has vigorous health as he supplies the conditions to produce it. He declines in health and dies suddenly or by slow degrees as he supplies the conditions that produce these results. It is all as certain as the rising of the Sun. No speculation, no guesswork. It is Cosmic Science.

Man breathes to live and he breathes to die. Inspiration may carry into this body either Life or Death. Which will it be? He is the master of his destiny. It depends upon the KIND of air one breathes.

Inspiration carries Life into the body when the air inhaled is fresh and pure. It carries Death into the body when it is vitiated. If the inhaled air contains enough poison, death comes quickly; and it comes by degrees, by installments, when the air contains less poison. A thousand persons died in less than two minutes when they inhaled the poisoned air in Hitler's gas chambers in World War II. Twenty million people died in the influenza epidemic of 1918-1919 from the poisonous gases of World War I which the winds curled around the earth. Millions all over the world are constantly dying by slow degrees from the effects of the vitiated air they breathe. The Breath of Life is the fresh, clean, outside air charged with Cosmic Force, to which the birds and beasts live in health and vigor. The Breath of Death is the foul, stagnant, polluted air of civilization, in which civilized man lives in sickness and misery, aches and pains, and grows feeble and decrepit when he should be in his prime.

Poisons In The Blood

Every person in a room needs 3,000 cubic feet of fresh air an hour. An adult poisons nearly a barrelful of air at each exhalation. The poisonous gases are brought by the blood to the lungs and eliminated. They consist of carbonic, lactic, hydrochloric phosphoric and other acids. You have heard of arterial blood and venous blood, but probably never knew the difference between them. Arterial Blood is bright scarlet blood that flows from the lungs back to the heart and then out over the body. This blood goes from the heart to the lungs as a dark, purple stream of venous blood that has returned to the heart from all parts of the body and was loaded with filth and poison. This polluted stream becomes bright scarlet arterial blood as it passes through the lungs and is purged of its cargo of pollution. Cell function in the body liberates into the fluid medium large quantities of these poisonous acids. Each cell must receive a volume of fluid equal to 2,000 times its own volume, and a volume of gaseous substance at least 20,000 times its own volume, in order not to be seriously poisoned within a few days by the poisonous acids in the blood.

This explains why air and water are so important in the body's vital economy, and why man dies by degrees from inhaling polluted air that fills his body with ailments, the symptoms of which are given names and termed 'diseases.' It is the marvelous perfection of the Blood Vascular System that enables the body to live with a volume of fluid hardly equal to one-tenth of its own weight. The speed of the circulation is sufficiently swift to prevent the composition of the blood, under normal conditions from being modified by the products of cell function. But that composition is seriously modified by polluted air, the bad liquids that man drinks and the bad foods that he eats.

Deadly Carbon Dioxide

During its passage through the lungs, the blood disposes of carbonic acid chiefly. This is the most common of deadly gases in the air of homes and hospitals, and is seldom seriously considered. This gas has the distinction of killing quicker than any other poison. Quicker than the venom of a rattler. That is the deadly character of the gas exhaled as the Breath of Death. It saturates the air of homes in winter when cold weather makes adequate ventilation impracticable. It is breathed over and over again, poisoning the body through and through, causing the members of the home to suffer from many ailments, including coughs, colds, sore throat, diphtheria, whooping cough, mumps, measles, scarlet fever, hay fever small pox, influenza, pneumonia. Etc. It is claimed these conditions are contagious. That is erroneous. Many people have them at practically the same time because they breathe the same kind of air.

Carbon dioxide is more dangerous because its presence cannot be detected by the five senses. It is colorless, odorless and tasteless. Combined with hydrogen gas, it forms the common fire-damp that sends many a brave miner to his death, and is the most feared of all underground enemies. Carbon dioxide is composed of one part carbon and two of oxygen in bulk, but by weight the gas holds 12 parts of carbon and 32 parts of oxygen. Both these gases are necessary to sustain the living organism, but in the wrong combination they are deadly enemies.

The atmosphere contains about one part of carbon dioxide to 2500 parts of air — a very small proportion. But this gas has a tendency to sink to the ground in low places. When there are three parts of carbon dioxide in 100 parts of air, a drowsy feeling appears. This can be relieved only by fresh air. The average person knows nothing about that. He falls asleep in this polluted air, and if he fails to wake up, it is called a "heart attack." When there are four parts of carbon dioxide in 100 parts of air, it is a fatal poison. When present in larger proportion, it is quick in its deadly effect and leaves no hope for aid or recovery. This gas sinks to the ground, and is sometimes found in large quantities in wells sunk in marshes and low lands. One author says: "A man went into a well in sight of his family. He failed to respond to a call, and they found him dead. His demise had been instantaneous. Thousands of such cases have occurred and are occurring."

The gas in sewers is also due to the presence of this poison. A man went through a manhole into a sewer only a few feet below ground level. Not returning in due time, a companion went after him. As the second failed to return, a third started to enter, but was stopped by the fourth. The first two were found dead, having died instantly by inhaling carbon dioxide. All the blood in the body passes through the lungs many times each hour, eliminating carbon dioxide gas and absorbing, in the lungs, the oxygen needed by the cells, and without which death comes quickly.

When not promptly eliminated from the body, carbon dioxide leaves a trail of damage in its course through the organism. It affects every cell, and as the cell is weakened the whole body suffers. Carbon dioxide is present in all charged drinks, in all soda waters, all beverages of the soda sort, in beer and fermented liquids, in cake, bread, baking-powder cookery, self-rising flour products, yeast bread, and in all fermenting products.

Exhaled Air Is Poisonous

At each exhalation the lungs throw out enough toxic gases to poison a barrel-full of air. The amount of poison eliminated by the lungs in 24 hours as carbon dioxide is equal to a lump of charcoal weighing eight ounces. That poison goes back into the body when we inhale what we exhale or what others exhale, as is the case where several occupy a room not adequately ventilated. When this exhaled air is breathed again and again, as it is in homes and hospitals, especially in winter when cold weather makes proper ventilation impracticable, the proportion of carbon dioxide and organic matter in it increases until it grows more dangerous to breathe. That is the principal reason why patients in hospitals develop influenza or pneumonia especially after operations. Their blood is poisoned by the anesthetics administered to dull the nerves enough so the body is insensible to pain, and, in addition to this poisoning process, comes the carbon dioxide in the hospital air the victim breathes. Lucky is one to get out alive.

The early symptoms of mild carbon dioxide poisoning are sensations of uneasiness, drowsiness, sneezing, languor, headache, sensation of oppression, coughs and colds. What fools one is the fact that the body, after a time, adjusts itself to a very vitiated atmosphere, and one soon comes to breathe, without sensible discomfort, an atmosphere which, when one first enters it, seems intolerable. This process of adaptation medical art terms "immunity." According to this theory, man becomes immune to a condition or a poison that fails to kill him on the spot. Such adaptation can occur only at the expense of a general depression of all the vital functions, which must be injurious if long continued or often repeated. In this condition people die by inches while being treated for some 'strange disease.'

The body is equipped with powers of adaptation that enable it to tolerate for a time an atmosphere so poisonous that it would kill a vital man in a few minutes if he suddenly walked into it. That makes it dangerous for a healthy man to breathe polluted air in smoke-filled rooms where half-dead men meet, play cards, etc. and do not seem to mind it. This little-understood power of adaptation of the living organism to its environment is well illustrated by an experiment of Claude Bernard, as has been stated. He showed that if a bird be placed under a bell-glass of such size that the bird will live for three hours, and is removed at the end of the second hour when it could have survived another hour, and a fresh, healthy bird be put in its place, the latter will die instantly. That is the fate suffered by the healthy man who tries to breathe the polluted air of the smoke-filled room where half dead men notice nothing.

The vital body does not resist the dangers of its environment. That is a false theory. The weak body tolerates thorn because it lacks the vitality to protest. Bernard demonstrated the body's power of adaptation. That is the power that enables the poisoned body of civilized man to drag out a miserable existence of 50 or 60 year in a polluted environment that would quickly kill a vigorous Indian brought in from the pure air of his forest home and thrust into that polluted environment.

Poisons Entering The Body

The effect of poison on the body depends on how it enters the body. There is a vast difference between poison entering the body through the lungs and through the mouth and stomach. Poison entering the body through the stomach does not contact the vital organs until it first passes through all the blood-making organs, of which the liver is the chief one, for which reason it is the largest of all the glands. The venom of a rattler is very deadly if injected directly into the blood, as when one is bitten by the reptile. The poison, if taken into the stomach through the mouth, would be neutralized so fully by the action of the fluids of the blood-making organs, and the refining and renovating processes of the glands of the blood vascular system, that when the poison reached the general circulation of the blood, it would be rendered so innocuous as to cause little more than slight illness.

When poisonous gases, acids and fumes enter the lungs with the air, they meet nothing to neutralize them. God never intended that man should live and labor in air so foul, that it would require a process of renovating, refining before being fit to breathe. So when poisons enter the lungs with and in the air, they pass directly into the blood, and may even cause sudden death, as they often do.

Big Battle For Health

The big battle for health in civilization is the struggle for good air. The body is not equipped to handle and neutralize poisoned air. This makes such air exceedingly dangerous. That is why one dies quickly in a closed garage filled with motor exhaust gas. That is why people are continually suffering with headaches and pains all over the body, in muscles, bones and joints. It is a case of blood poison and much of the poison enters the blood through the lungs.

One's blood must be poisoned before one can be ill. The easiest and quickest way to poison the blood is to inhale polluted air. Children in cold regions must remain indoors so much in winter, that they are sick with all kinds of ailments from coughs, colds, mumps and measles, on to the more serious conditions of whooping cough, tonsillitis, diphtheria, scarlet fever, asthma, influenza, pneumonia, etc. The air of your environment, your home, the place where you labor, where you live and sleep, is constantly saturated with a hundred poisons, and anything can happen to you by inhaling that horrible air. You have a case of blood poison from which you may suffer with any ailment.

If the condition is not serious enough to kill you instantly, you live; but you are certain to suffer from time to time in time in some way, from a cough and cold to the more serious states. It is easy for you to test a serious case of blood poison. Enter your garage, close the door and all ports of ventilation, then start the motor of your car and see what happens. You will soon faint and fall to the floor, yet suffer no pain; and that will be your end unless help quickly comes.

Now for a milder case: You live in the cold region where doors and windows of the hone are kept closed in winter to keep out the killing cold. In that home the air is unfit to breathe. First it is poisoned by the fumes of your own lungs; then the fumes of the cook-stove and heater; then the fumes of cigars, cigarettes, pipes, etc.

One hundred times each hour every drop of your blood is sprinkled through that poisoned air in your lungs in a shower of Red Mist. Your lungs are filled with that polluted air; your blood becomes saturated with poisons. The surprise is that the blood holds up as well as it does under such blood-poisoning conditions. If the air were sufficiently polluted, you would faint and die as you would in your garage. If it were that serious, something would be done to improve the situation. Yet helpless infants die in their sleep because of the foul air of the home. Adults come down with sickness, and they are told it is the work of germs.

Millions Chronically Ill

The press of September 7th, 1951, reported a statement of Dr. A. C. Knudson, chief of the Veterans Administration's physical medicine and rehabilitation division made to the 29th annual session of the American Congress of Physical Medicine; at Denver, Colorado, that 25,000,000 people in the U.S.A. are chronically ill, *"and warned that their number is annually increasing."* Included in Knudson's list were 1,000,000 persons paralyzed on one side; 2,500,000 orthopedically disabled; 1,000,000 diabetics and about 10.000,000 afflicted "with disease of the heart and arteries." In the face of this horrible record we are constantly told in the big publications, of the "great strides being made by medical science."

Lesson No. 26
Poisoned Air

"Human experience with the poisonous effects of carbon monoxide gas probably had its beginning in the prehistoric ages.... Dr. L. Lewin, who states that his report on the history of carbon monoxide poisoning is the first of its kind, has traced references to the action of this gas back through the ancient Greek and Latin literature and concludes that this poisoning 'of all stands alone in its close relation to the history of the civilization of mankind'." — Review of Carbon Monoxide Poisoning, 1938, by R. R. Sayers, Senior Surgeon, U. S. Public Health Service.

It seems strange that this matter should be so late in receiving attention, when one author says the Carbon Monoxide Gas is the deadly agent, the great killer, the leading life destroyer. When man made his first fire, he thought he had something and little dreamed that he was setting into motion the production of a destructive gas that has killed millions, and will continue to kill millions until the present arrogant civilization has disappeared. The first materials used for making fire were grasses, wood, and other vegetal matter. Ancient records show that many cases of fatal poisoning followed from the fumes of fires. It would seem that early men were unaware of the dangers lurking in the poisonous gases set free by fire. Modern man knows little more after living in the midst of these gases for thousands of years. He knows not that carbon monoxide kills thousands quickly and millions by slow degree. Physiologists declare that progressive damage occurs to the body from the inhalation of any gas so deadly as to cause death under certain circumstances. The gas may be so weak as not to cause sudden death, but constant contact with it induces a deteriorative process in the organism, the destructive work of which appears in various symptoms which are called "disease."

It was not until Gustavus Magnus in 1837 proved the presence of the "blood gases" in different proportions in the blood, that the present theory of respiration assumed definite form. That cast some light on the secret of respiration and animation. Little is known about the function of respiration — yet it is the primary process of living. The object of all other functions is to keep the breathing organs in condition to perform their work properly. When they fail, death ensues.

Prof. Hotema   *Man's Higher Consciousness*   *Lesson No.26*

Carbon Monoxide Gas

From birth to death man must breathe constantly to live. When he stops breathing he stops living. Naturalists use this evidence to show that the most dangerous substance to health and life is polluted air. Yet this important branch of knowledge has been considered so lightly that little literature bearing on it has been produced. When Dr. Sayers wrote his "Review of Carbon Monoxide Poisoning" in 1936, it soon went out of print because it covered a field in which there is no use for vaccines and serums. It was not until 1920 that Dr. L. Lewin published in Berlin his work of 369 pages on the dangers of Carbon Monoxide Gas, and in it he stated that his report was the first of its kind ever written.

Lewin found carbon monoxide gases mentioned in ancient literature, quotations from which show that this poison was a frequent cause of death by accident, by suicide, and by use as an agent of punishment and torture. He quotes a statement from Livius that during the second Punic War, about 200 B. C. — "The commanders of the allies and other Roman citizens were suddenly seized and fastened in the public baths for guarding, where the glowing fire and heat took away their breath and they died in a horrible manner."

Julian the Apostate (331-363 A. D.), tells how he was almost suffocated while in winter quarters in Paris. Because of cold, he had a small fire in his room. The fumes from it affected his brain, put him to sleep, and he was carried out unconscious. Otherwise he had perished. Campegius, who lived in the 15th century, told of two merchants, travelling toward Lyon in winter, who stopped at an inn for the night and, to warm the room, made a fire in the fireplace and went to bed. Next morning they were found dead in bed.

After the 15th century, with the increased use of coal as fuel, poisoning by the gas greatly multiplied. With the inventions of methods of producing heat for homes and industrial use, dangers of poisoning by the gas have increased at an alarming rate, until today, with many additional hazards from the wide use of gas-burning appliances and manufactured gas containing large amounts of carbon monoxide, this form of poisoning has become one of the most widely distributed and most frequent causes of accidents and death.

Kober and Hayhurst investigated the matter and reported a list of 24 possible sources of contact with carbon monoxide gas in industrial operations alone. The increasing use of motor cars, trucks, buses, and other gasoline-burning engines has made exhaust gases a constantly increasing source of carbon monoxide poisoning. The dangers have been increased by adding certain chemicals to the fuel to prevent accumulation of carbon in the engine. These chemicals in the exhaust gases are more poisonous than the fumes of the gasoline. The fact that no noticeable odor warns one of the dangers from carbon monoxide gas, was first mentioned by Baconis de Verulamio in 1684, and, unlike most of his predecessors, he was careful to mention "vapor carbonum" instead of "fumes." Van Halmont was the first investigator to term such fumes "carbon gas." It was not until 1732 that Boerhave made what is probably the first animal experiments with carbon monoxide gas. He found that all red-hot matter, as wood and coal, produces a vapor so fatal that it quickly killed an animal shut up in a confined space.

It is terrible to think what that vapor does to people in cold regions who live all winter in closed quarters, with little or no ventilation, and breathe that vapor for weeks and weeks. Their spiritual (air) organs are damaged beyond repair. In 1919 the Bureau of Mines published a technical paper on the results of studies made of the degree of vitiation of garage air by motor car exhaust gases, in which it was said: "In tests made by the authors, garage air was rendered decidedly dangerous after an automobile engine had been running ten minutes."

Henderson and Haggard report that when a motor car is running ten miles an hour, occupants of a car 40 feet behind are surrounded by exhaust gases diluted to a concentration of one or two parts of carbon monoxide to 10,000 parts of air. They further state that one part in 10,000 is a frequent condition of the air in city streets where traffic is heavy, and increases as traffic increases. In 1920 certain research workers made an investigation in which 1308 garage and repair shops, 341 in New York City and 967 in the rest of the state, were visited. These shops employed 5908 men.

Dangers Of Poisoned Air Unknown

Most of the men in these garages and shops were totally ignorant of the dangerous properties of the exhaust gas of cars and trucks. Some did know it contains "knock-out" properties, but knew not that serious and even fatal results would follow its inhalation. Others believed that they acquired immunity and could not be injured after working in a garage a certain period. People know not that they are constantly surrounded by an unseen foe to health and life. They know not that most cases of sudden death from so-called heart attack are the work of this unseen foe.

They know not that in a certain area that is free of this unseen foe, people live to be 200 and 250 years old. Inform yourself before it is too late and avoid this dangerous enemy. Study this picture and see how that unseen foe, in the form of invisible gas, enters nose and mouth and goes directly into the lungs, where there is nothing to prevent the poisonous gases from passing directly into the blood and being carried into the deepest recesses of the body.

In experiments at the Bureau of Mines, dogs became unconscious after a motor was run ten minutes in a single car garage with the door and window closed. One dog died in 30 minutes and the carbon monoxide concentration was only 1.5 percent. Acute cases of poisoning by carbon monoxide gas result in death by asphyxiation through its deadly action of: 1. paralyzing the nerves of the breathing centers of the brain, and of 2. changing oxygen-carrying hemoglobin into non-oxygen carrying carbon monoxide hemoglobin.

Chronic cases of poisoning by the gas result in a lingering death, in which the victims, before they die, suffer sometimes for years, while their poisoned bodies present various symptoms resulting from the destructive work of the gas. The symptoms are named and treated as "diseases," while no effort is made to locate the cause. While billions of the tax-payers money are spent annually for the alleged improvement of the public health, under the direction of political health officials, this terrible menace to health and life marches on unmolested, striking down the millions in its path.

Brain Poison

As the mind grows feeble, man's physical world grows smaller. Civilized man has lost contact with the Spiritual World because he has lost his Spiritual Powers, and he loses contact with the physical world as he loses his physical intelligence because of damage to his brain.

The press of October 13th, 1947, contained an item headed, "Insanity More Prevalent in City," which said: "The nearer you live to the center of a large city, the more likely you are to go insane. These are the conclusions of a study of the geography of insanity in five large cities of the U.S.A. Psychiatrists have long known that city people go crazy more often than country people do, but the discovery of well-defined insanity zones within cities surprised even them. The rate of lunacy lessens as you travel from the center of a city."

The highest incidence of insanity occurs in the center of cities where the air is more heavily charged with carbon monoxide gas from motor cars, trucks, coal smoke, tobacco smoke, etc. As we move out toward the periphery of the cities, there is less carbon monoxide in the air and less insanity among the people. As we reach the better air of the open country, the incidence of insanity diminishes still more, and perhaps would disappear entirely if country folks never visited the cities, and if their homes were not frequently filled with tobacco smoke, fumes of cookery, of the heater and cook-stove, and perhaps of an oil-burning lamp. Certain so-called diseases are but the symptoms of slow carbon monoxide asphyxiation. These symptoms are chiefly: Headache, dizziness, nervousness, nerve and muscle pains, digestive disturbances, restlessness, weakness, impairment of vision and hearing, shortness of breath, anemia, hyperemia, angina pectoris. Frequently the appearance of the victim is that of one drunk. The eyes may appear dull, more or less fixed, and somewhat bulging. The order of the respiration changes — the rate is first increased, and later slowed and irregular. When birds are exposed to carbon monoxide gas they appear moribund. They show signs of distress when exposed for an hour to air containing 0.1 percent of carbon monoxide, and within two to five minutes when exposed to air containing 0.2 percent.

Pernicious Anemia

Koren showed that progressive pernicious anemia is a symptom of carbon monoxide poisoning, and he described the following pathological effects: Dilation of the heart, enlargement of the spleen, large decrease in the number of red blood corpuscles, and peptonuria.

Postmortem of a fatal case showed that all internal organs exhibited great pallor; the heart muscles were thickened; the heart was microscopically yellow dotted and showed advanced fatty degeneration; spleen was considerably enlarged and of hard consistency. There are various causes of hardening of the blood vessels, and one of these is carbon monoxide poisoning. Mott made a postmortem examination in the case of a woman found unconscious from illuminating gas poisoning and who four days later without regaining consciousness, died. He said that he never saw such extensive and general capillary hemorrhage in the brain as in this case. Pulvertaft reported a case of spontaneous rupture of the heart of a youth of 19 due to carbon monoxide poisoning. Lewin found that carbon monoxide destroyed brain function of a dog so it did not know its master. He stated that in cases of carbon monoxide poisoning, great changes of deterioration in the brain occur sooner or later. Symptoms of paralysis and other nerve disorders present in cases of carbon monoxide poisoning, show the specific effect of the poison of the brain and other nerve centers. For this reason, carbon monoxide gas is called a brain poison. Wendell Winkle, Republican candidate for president in 1940, died in December 1944 in his sleep in a certain hospital. The report said, "His wife, standing by, looked down into his still boyish face as his life flickered out." Basic cause of death, carbon monoxide gas.

On March 13th, 1943, J. P. Morgan, noted New York banker, "who made his banking firm a colossus of the financial world and his very name a symbol of extreme wealth and power," died of "a heart ailment." Basic cause of death, carbon monoxide gas. In February 1945 Chief Justice Edward C. Eicher of the U. S. District Court, "died in his sleep at his home, age 65." Basic cause of death, carbon monoxide gas.

December 5th, 1944, Roger Bresnahan, considered by many as one of the greatest major league catchers of all-time, died of "a heart attack" He was 64. Basic cause of death, carbon monoxide gas.

Cerebral Hemorrhage

Cerebral hemorrhage caused the death of General Edwin M. Watson on February 27th, 1945, age 61. He was the late President Roosevelt's military aide and made his official appointments. Basic cause of death, carbon monoxide gas. In April, 1945, President F. D. Roosevelt died suddenly of cerebral hemorrhage, age 63. Basic cause of death, carbon monoxide gas. The press of January 31st, 1948, reported the death of Herb Pennock, one time star baseball pitcher, age 53. He died "after a cerebral hemorrhage." This man "collapsed as he entered the

Waldorf-Astoria Hotel (in New York City) to attend a National League meeting. There was no hint that he was ailing. A few hours before his death he had invited friends to attend the fights at Madison Square Garden tonight," said the report.

Lung Cancer

Dr. T. R. Van Dellen, who writes a daily column on human ailments, says that 62 percent of the cobalt miners in Schneeberg, Germany, die of lung cancer, and adds, "A similar catastrophe occurred among the neighboring pitchblend miners of Joachmsthal." The cause is poisoned air. Chemists, who analyze the chemical content of the air of some of our big cities, found that there is a very high content of different chemical byproducts definitely harmful to man. The list included some 27 poisonous byproducts in every cubic centimeter of air. The press of August 7th, 1945, stated the gun crews on war ships, "choking, gasping, wiping their streaming eyes, keep feeding the guns," until they collapse into unconsciousness from the fumes and smoke that fill the turrets, and must be carried out. The press of March 14th, 1946, stated that acid soot falling in some of the large cities was so strong, that when it lodged on the nylon stockings of the women "it ate holes right through them."

What does that acid soot do to the delicate lining of the nose, sinuses, trachea, bronchi, and cells of the lungs. And each city is trying for more factories, which means more acid soot for those in the cities to inhale. Professor H. Landsburg, Geophysical Laboratory, Pennsylvania State College, stated that wherever human dwellings are, wherever industry has found a foothold, the air is polluted with poisonous fumes and gases. He said: "Among the more dangerous compounds in the air, nitric acid and sulphuric acid are always present in combustion gases. Sulphuric acid fumes, being heavier than air, float like a death-pall over large cities, and are so corrosive that they injure everything with which they come in contact. Man's breathing organs are consumed by the corrosive action of the acids, causing his voice to weaken and sometimes it fails entirely."

The rapidly increasing mental weakness of the people in the U.S.A. is startling evidence of the deadly effect of carbon monoxide gas. The press of November 28th 1947, stated the psychiatrists estimate that 1 in 16 in the U.S.A. is mentally weak. Damage to the brain caused by poisoned air was studied by Dr. John Chomyak and Dr. R. R. Sayers of the U. S. Public Health Service. They examined under microscopes the brains of four dogs, each killed by breathing for less than thirty minutes a small percentage of carbon monoxide gas in motor car exhaust fumes.

Nerve Cells Destroyed

They found that nerve-cells of some of the most vital parts of the brain were almost entirely destroyed. Some cells had ruptured (cerebral hemorrhage), and were partly liquefied. Others were shrunken and distorted. Blood vessels in the dogs' brains were swollen and clogged with stagnant red blood corpuscles, as the body forces tried to give aid by rushing larger supplies of life-sustaining oxygen to the damaged brain-cells. These doctors' findings showed that "cerebral hemorrhage" results from a rush of blood to the endangered area, as the body's vital force tries to carry more oxygen to the brain-cells. An item in the press of September 24th, 1930, stated, "Eventful death of all plant-life in America's big cities is certain unless smoke and exhaust fumes of motor cars, trucks and buses are curbed, was predicted by Dr. D. S. Johnson, director of the botanical gardens of Johns Hopkins University, Baltimore.

Black Lungs

An item in the press of November 13th, 1938, headed Cleveland, said: "Fifty thousand tons of soot, tar and other filth float in the air of this city of a million people — 100 pounds for each person. One of the dirtiest sections produced 87.15 tons of grime and insoluble solids, such as carbons, tar, fly ash and ferrous oxide."

City air is a poisonous mixture of smoke, soot, fumes and acids, which include such poisons as carbon monoxide, carbon dioxide, sulphuric acid, hydrochloric acid, nitric acid, hydrocyanic acid, benzene, methane, and other dangerous chemicals. In addition to these poisons, city air is saturated with the exhaust fumes of motor cars, trucks, buses, gas engines, etc. These exhaust fumes consist of carbon monoxide, carbon dioxide, lead oxide, lead carbonates, free gasoline and complicated benzene chain compounds of the hydrocarbon series. Tasteless, colorless, odorless, invisible, carbon monoxide gas takes a terrible toll of human life in the cities. The larger cities have a huge, dark, gas blanket hovering over them that holds down the gases and tends to smother those living and laboring in the cities. Writing on Our Smoky Cities" in Collier's Weekly, W. B. Courtney said: "As you fly across the U.S.A. you see a chain of dark smudges on the skyline, like blots from a cosmic thumb. Those are the cities. Certain pilots in daytime will name the cities without consulting maps. They do it by the size of the black umbrella that hangs in the air above it. Sometimes the larger cities raise smoke umbrellas with 200 square miles of spread. I have looked at Chicago, St. Louis and Kansas City, among other cities, on clear days, and seen nothing but a pall of smoke and soot. The disturbing thought is that under those black, poisonous umbrellas millions of people live, labor, sleep, seek health, happiness and fortunes; that millions of children struggle under those poisonous umbrellas for a chance to grow into sound maturity and optimistic citizenhood."

The Cincinnati Post of April 1946, stated that the fall of soot and ash on that city's 73.9 square miles during March amounted to 2725 tons or 227 railway carloads. It was sufficient to have covered a 40 x 150 feet lot to a depth of more than 75 feet. During the year 1945 Cincinnati was deluged with 33,231 tons of soot and ash; and air pollution in that city is no worse than in other cities of similar size. Some of the disaster that comes to those living under these poisonous umbrellas was contained in a report made in 1931 of a two-year survey by the Mellon Institute in Pittsburgh. In part that report said: "Constant inhalation of poison-ladened air results in a gradual process of absorption of the body of the poisonous products of combustion. This insensible intake may cause any acute disorder. The process of slow-planning eats away insidiously at the vital tissues, making it impossible for the body and brain to function properly."

Investigations show that the lungs of those living in the larger cities become black as coal. Dr. Thomas Darlington, former health commissioner of New York, wrote: "I have performed many autopsies upon New Yorkers, and almost without exception their lungs were as black as night." Harold D. Blackwell, educational director of the Smoke Prevention Association, in an address in Milwaukee, said: "The lungs of anyone living in Milwaukee for five years, become as black as coal; but if that person lived in the country where the air is better, his lungs would be pink and grey, the natural, healthy color."

This knowledge reveals the secret why people become short-winded as they grow older. The polluted air causes the walls of their

Prof. Hotema   *Man's Higher Consciousness*   *Lesson No.26*

lungs to thicken, and the walls to become coated with carbon, making it difficult for the gases of the air, needed by the body, to pass through the lung-walls into the blood. The result is labored breathing, especially on a little exertion. Authorities have demonstrated a concentration of 0.62 parts of carbon monoxide per 10,000 cubic centimeters of air at street level in busy sections of cities of 500,000 population and over.

    Few poisons in the air are more deadly than carbon monoxide. Air containing as little as one-twentieth of one percent will cause headache, and one-fiftieth of one per cent may cause total collapse. Dr. L. Burns examined blood specimens of more than 20,000 persons to discover the affect of carbon monoxide on the body, and wrote: "Carbon monoxide gas seeps into the blood through the lungs, and mixes with the hemoglobin to such extent that the blood cannot perform its normal function of carrying oxygen to the rest of the body."

    This gas seeps into the blood and it is absorbed by the hemoglobin, whose normal function is to carry oxygen to the cells. The hemoglobin has an affinity for this gas about 300 times greater than for oxygen, making very rapid the absorption by the blood of this gas. As the hemoglobin becomes saturated with carbon monoxide, the oxygen in the blood is reduced in proportion. The first symptoms are headache and weakness. More serious symptoms soon appear as the condition progresses. Scientists of Harvard Laboratories, risking their lives to learn more about the symptoms of carbon monoxide poisoning, found the average man can endure it only until his blood is one third saturated. The serious danger of the gas was shown by the way it affected one of the scientists. He had just finished some tests requiring great skill and was feeling no ill effects from the gas, when he suddenly collapsed and had to be carried out in the fresh air and revived. Small concentrations of the gas soon bring man down to the breaking point. Five percent of the motor cars and closed trucks on the highways have sufficient concentrations of gas to be a menace to the drivers and passengers. Only one part of the gas in 1000 parts of air can render a man unconscious if he inhales it for 30 minutes. There is no natural or acquired immunity to the gas. Repeated exposures produce the same effect each time.

    Hydrocyanic acid gas is worse. Only a few grains of it produce violent death. Most people who drop dead or die suddenly are not afflicted with heart ailments as is claimed. It is the work of polluted air.

Causes Cancer

In the press of August 19th, 1932, appeared excerpts from "the annual report of the Bernard Free Skin and Cancer Hospital," in which it was asserted "that city dwellers, breathing polluted air, develop cancer of the lungs at a rate three times greater than inhabitants of rural districts."

The press of October 24th, 1936, stated that evidence showed an increased occurrence of cancer of the eyes "resulting from colds," according to the doctors. The report was made "at the closing session of the clinical congress of the American College of Surgeons." Quite an imposing body, but none of those present could offer any suggestion as to why an increased occurrence of cancer of the eyes should appear as the result of colds. The Mellon Institute of Pittsburgh issued a report in 1931 of a two-year survey covering the effect of polluted air on human health. In part the report said: "The inhalation of poison-laden air results in a gradual process of absorption by the human body of the poisonous products in the air. The effect of this insensible intake is cumulative and results in a condition of slow poisoning that insidiously eats away at the vital tissues of the body like cancer."

The Chicago Health Department reported that in Chicago there is so much sulphuric acid gas in the air that it rots clothes hung on wash lines, and eats away at building stone and metal guttering. The report stated that while copper guttering in rural regions beyond the smoke zone lasts almost indefinitely, in the larger cities it is destroyed in about ten years by the corrosive action of polluted air. Can flesh and blood endure long under a condition that "eats away at building stone and metal guttering?" Think of babies and little children that must breathe that deadly air.

Eat Up The Body

Man does well to survive, for thirty or forty years in air saturated with poisons and acids so destructive that they eat up clothing, copper guttering, stone and steel monuments. The corrosive acids in city air attack the cells, tissues, throat, nose, lungs, and brain, and all organs and glands of the body. They attack the blood corpuscles and cripple them so they cannot carry on their normal function. That condition is termed anemia. They attack the nerves, and the resulting pains are called neuritis. As the nerves weaken, paralysis may result, and often does.

They attack the muscles, producing dull pains that puzzle us, this is called rheumatism or lumbago. They attack the joints, and this is called arthritis. They attack the air chambers of the head, and it is called sinusitis. They attack the throat and it is termed laryngitis, tonsillitis, diphtheria, etc. Hoarseness often follows, and in time one's voice weakens, or may be lost. They attack the muscles and nerves of the heart, and it is called heart disease. They attack the lungs and it is called hay fever, asthma, or tuberculosis. They attack the pancreas, and it is called diabetes. They attack all parts of the body. Medical names, names mean nothing except to indicate that part of the body where degeneration is most serious from the evil work of polluted air.

The press of October 24th, 1936, quoted a report made "at a closed session of the Clinical Congress of American College of surgeons" The report showed that increase in cancer appeared in patients following recovery from influenza, in workmen handling oil substances, such as garage men, mechanics, oil station attendants, and auto salesmen. All the doctors present were puzzled. They had no answer, no suggestions. Not one of them had the slightest suspicion that polluted air had anything to do with the matter. It was said that after influenza epidemics, as that of 1918-1919, a threefold increase in cancer of the eyes was found. Dr. L. A. Lane rose up and said: "Not a few patients date the beginning of tumor from an attack of influenza, pneumonia, or a severe cold."

All these disorders, including the cancerous conditions that follow as a sequel, are the evil work of polluted air and the drug poisons which are used to treat the patients.

Smoke, Soot, Tar, Acid, Gas

Soot is a mixture of carbon, ash, tar, sulphuric acid and other poisonous gases. In the industrial cities the soot-fall amounts to hundreds of tons per square mile a year. This soot-deposit contains as much as several percent of tarry matter and 20 to 30 percent of carbon. Both substances are active in the aging process of hardening the body and stiffening the joints. The constant fall of soot covers everything. The interior walls of the lungs become coated with the tar and carbon, making breathing hard and preventing the passage of oxygen into the blood.

H. B. Meller, of the Mellon Institute of Industrial Research, said: "When it is known that one takes about 30 cubic inches of air into one's lungs in each inhalation, or about seven times the weight of food and water consumed, it can be understood why more people are weakened, devitalized and poisoned by the pollution in the air they suck into their lungs, than by all the ingredients in the food they eat and the water they drink." Man longs for health and spends much money trying to gain it, yet with every breath he fills his lungs and blood with the Agents of Death. No one living in the cities can escape it. Scientific investigation shows that city air is a poisonous mixture of industrial fumes, such as carbon monoxide, sulphuric acid, hydrochloric acid, nitric acid, hydrocyanic acid, benzene, methane, sulphur, and other deadly chemicals too numerous to mention. Sulphuric acid gas is heavier than air, and hangs like a death-pall in and over the cities. This gas is so corrosive that in certain sections the fumes eat clothing hung on wash lines. It eats ulcers in the skin, it consumes the lungs of those who breathe it.

Dr. Darlington says: "The products of combustion irritate the eyes, ears, nose, throat, the respiratory tract, the bronchial tubes, and the gastro-intestinal areas. In the lungs the carbon particles accumulate and become imbedded in the air cells, and in time the lungs change from natural pink to black." As these poisons filter into the blood through the lungs, the body must take vigorous measures to eliminate them to save life. One of these is through the skin by means of a heavy rash called small-pox. If people knew this and would give smallpox patients fresh air, they would soon recover. Professor Godfrey Rodriguez, in his work on air termed "The Key To Life," wrote: "The most forceful proof of the power of air was demonstrated in London in 1912, when 150 smallpox patients were taken into a field because the hospital was on fire. They had to spend three days and nights exposed to all sorts of weather, but they breathed good air and all recovered. In Glasgow, Scotland, in 1914, when ventilation was introduced in a certain block of buildings, in eight years thereafter only four cases of typhus occurred, in contrast to 107 cases in the same block of buildings in a single previous year." Of course some people are not pleased to have such information leak out to the public.

Tobacco Smoke

The U.SA. in this generation has developed into a tobacco-saturated nation, thanks to the diligent work of the tobacco trust, which uses the doctors to promote the sale of its products. The doctors live in glass-houses and cannot afford to throw stones. They have studied the toxicology of poisons and know the evils of using tobacco for chewing or smoking. They cannot warn the public of the poisons contained in tobacco, because medical art considers the more virulent poisons as the best remedies. Nicotine is an alkaloid and a narcotic. It is deadly dangerous. Just two drops of it will kill a man; 8 drops will kill a horse, and 50 milligrams will kill a 20 pound dog. It takes 1,000 milligrams to make one gram and 30 grams to make one ounce. There are 150 to 400 milligrams of nicotine in three ounces of tobacco, and a smoker takes almost three milligrams of nicotine into his body every time he smokes one average cigarette. That is enough to kill a man instantly if all taken at one dose; and it kills the smoker by inches because the repeated small doses of nicotine are cumulative.

The poisons found in a chemical analysis of tobacco are as follows: "Nicotine, carbon monoxide, nictinuine, carbon dioxide, ammonia, methane, methylamine, hydrogen-sulphide, furfural, pyrrole, pyridine, picoline, lutidine, colloidin, formaldehyde, carbolic acid, prussic acid and arsenic." Chemical analysis shows that a cigarette contains: "Furfural, acrolein, diethylene, glycol, nicotine, pyridine, ammonia, carbolic acid, carbon monoxide, and a host of tarry substances."

The press of September 4th, 1947, reported Dr. A. C. Ivey, Vice President of the University of Illinois, as saying: "A person who smokes a pack of cigarettes a day, in ten years inhales eight quarts of carcinogenic tar substances into his lungs that are sufficient to produce cancer. As the inhaled tar substances enter the lungs, the first damage is done to the delicate air cells and their lining membrane. Non-smokers do not escape. They inhale tobacco smoke with every breath almost everywhere in the nation. Every public building, bus and train are filled today with tobacco smoke, including most of the hospitals, and with every breath one takes cancer-causing tar into the body."

Tobacco smoke is a cloud of tiny particles of exceedingly fine carbon-dust coating in the air. He who inhales that smoke-laden air will in time have a coating of carbon on the interior surface of his lungs, which prevents poisonous gases of the blood from being eliminated, and prevents the vitalizing gases of the air from passing into the blood.

Dr. A. H. Roffo made a special study of the matter and found that the carcinogenic action of tobacco tar upon the human body is more active and death-causing than coal tar. He found that benzpyrene, one of the constituents of tobacco abundantly produced by tobacco smoke, is a very virulent carcinogenic. In his report he said: "Due to the prevalence of smoking, although tobacco tar enters our bodies in tremendous volume, yet almost the entire population of the world is kept in ignorance by the tobacco manufacturers regarding the dangerous nature of the tar."

The students of a pharmacology class took the tobacco from two cigarettes and boiled it in a little water for a few minutes. They placed two drops of the brew on a cat's tongue, and within two minutes it was in convulsions. Another drop was placed on the cat's tongue and it quickly expired. Tobacco poison strikes first at the brain, causing mental confusion, giddiness, faulty memory, and general deterioration of the intellectual faculties. Tobacco poison constricts the arteries, causing blood pressure to rise and making the function of the heart more difficult.

The constriction of the blood vessels reduces blood supply to the cells, and may cause cramps in certain muscles. The condition sometimes becomes so serious that finally gangrene of the extremities develops. A test showed that a habitual smoker raised his blood pressure 25 points in 20 minutes by smoking three cigarettes, and it required an hour for it to return to normal.

Coronary Thrombosis

A serious condition may develop in the small arteries which supply the heart muscles, and coronary disorders strike man in his prime, causing cramps around the heart that may imperil life within a few minutes. Of a recent survey of 150 victims of coronary thrombosis, 94 percent were smokers. The remaining six percent had quit smoking only a short time before death. The press of August 7th, 1945, stated that U. S. Senator Hiram W. Johnson, long a leading light in California politics, died in a coma the day before, the cause of death being "thrombosis of a cerebral artery".

H. Wells, publisher of Health Culture, stated that he had tried to find a case of coronary thrombosis in which the victim was not subjected to the effects of tobacco smoke, and failed to find one. He wrote: "Two women I knew died recently of this trouble, and their husbands were both heavy smokers. The rooms they occupied were filled with tobacco smoke. The women might as well have been smokers so far as the effects of the smoke on them were concerned. One man I knew was begged by his relatives to stop smoking, and by his wife in particular. He had an attack of coronary thrombosis and died." — Health Culture.

Science Editor Dies Of "Heart Attack"

The press on May 3rd, 1952, reported the death of Howard W. Blakeslee, age 72, "of heart attack." He "was stricken at his home in Port Washington, New York, "with coronary thrombosis." For a quarter of a century, according to the account, the decedent had been—"Associated Press Science Editor and a pioneer in making science clear to the layman." How did he do it when scientists are so confused and puzzled about the Universe and its laws that their theories change overnight. The report stated that Blakeslee was "a co-winner of the Pulitzer Prize in 1937" and of "numerous other honors for his reporting in the field of science." Years of reporting and learning, continues the account, led him to this conclusion: "Science has given us more of everything, including more opportunity to develop morally and spiritually."

The real facts are that science has done nothing to develop man spirtually, from which it appears that Blakeslee did not understand what the term "spiritually" means. Science has done its utmost to obstruct spiritual development, and claims that materialism includes and encompasses everything, while spiritualism is only a heathenish superstition.

G. K. Chesterton saw the matter correctly when he wrote: "Man is the creature that progress professes to improve. . . . There has certainly been a rapid series of inventions; and, in one sense, the activity is marvelous and the rapidity might well look like magic. But it has been a rapidity in things going stale; a rush downhill to the flat and dreary world of the prosaic; a haste of marvelous things to lose their marvelous character; a deluge of wonders to destroy wonder. This may be the improvement of machinery, but it cannot possibly be the improvement of man. And since it is not the improvement of man, it cannot possibly be progress."

Cigarette Consumption

Since 1910 the consumption of cigarettes has increased nearly 500 percent, with a corresponding increase in ailments caused by tobacco poison. In May 1951 the tax collected by the State of Florida on cigarette sales amounted to $1,338,000.00. In the forty-eight States of the Union, on that basis, the tax on cigarette sales in one month would amount to $64,224,000.00. That is the rate the people of the U.S.A. spends money each month for just one method of poisoning their bodies, destroying their health and shortening their lives.

Dr. John A. Killian, head of the department of biochemistry of the Post Graduate Medical School of New York, found that smoking definitely increases the content of carbon monoxide gas in the blood. His investigation showed that with every puff of a cigarette, a portion of carbon monoxide, a deadly poison, enters the blood through the lungs.

Dr. C. Barber was reported in the press of September 23rd, 1927, as declaring before the American Association of Medico-Physical Research at Chicago, that 60 percent of the babies born of smoking mothers die before they are two years old. In part, he said: "What we breathe has much more to do with the action of the ductless glands, the functions of the organs, the nutrition of the body, and the development of the nerve system than what we eat and drink. When people breathe smoke-laden air, it leads to degeneration of the heart, liver, and other organs and glands" (N.Y. Times, September 23rd, 1927.)

The Aero Medical Association of the U.S.A. was told that carbon monoxide gas from burning cigarettes injures the sight of pilots in the Air Force. The doctors who made the investigation reported that inhaling the smoke of three cigarettes causes the loss of vision, which occurs at an altitude of about 8,000 feet.

King George VI

The King of England was in bad health. The great doctors of London examined him and said he was suffering from "structural changes of the lungs." Without rhyme or reason the tissues of the lungs insisted on changing and causing the King misery. The great doctors rose to the occasion and decided to show the unruly lungs a trick. So on September 23rd, 1951, they cut out one lung, or most of it, and cast it into the garbage can. Unless the remaining lung takes a hint and becomes obedient, it will get the same dose — and the grave will get the King's body. (PUBLISHERS NOTE — this was written by Klamonti shortly after this operation and time has shown his statement to be true.)

Poisoned air is no respecter of persons, caring not whether you are priest or pagan, king or common herd. For years we have written about the dangers of poisoned air, which in this age fills every city, home and hospital. As the poisoned air comes in contact with the air organs, the natural result is damage to the lung structure in the form of degeneration. The degeneration in lung tissue the doctors term "structural changes." That is exactly what it is. What they fail to realize is that "structural changes do not occur without cause, and that cause is not removed by cutting out the lungs. Had the doctors known about poisoned air and the damage it does to the body, they would have advised the King to go and live in the country, far away from the tobacco smoke, factory soot, motor car fumes, carbon monoxide gas and poisonous acids found in all city air. Then the forces of the body had repaired the lungs and in due time the King would have felt like a new man.

Change Your World

In Lesson No. 12, under DANGER OF ABRUPT CHANGES, you are warned of the danger involved in attempting to change too suddenly from one mode of living to another. Sudden shocks must generally be avoided. The body must have time to adjust itself to new conditions. But this does not apply to a change from bad to good air. Such change can be made any time, and life is often saved by making the change quickly. With these facts in mind, you will understand that if a 100 percent Breatharian suddenly walked into the poisoned air of our civilization, a few breaths of it would cause him to fall unconscious, because his body would not be adjusted by years of endurance to tolerate the poison. Then he would be hustled to a hospital, with its stale, polluted air, where he would be scientifically polished off in the "oxygen tent" and prepared for the cemetery.

We must recognize cosmic law in our desire for improvement or be disappointed with the body's reaction. It would be dangerous for one to strive for the perfect state of Breatharianism while living the conventional course in the poisoned air of our civilization. You cannot safely change to perfection from imperfection without first changing the environment in which you live and labor. You cannot keep what you have and have what you want. The Law of Compensation exacts a price for every privilege.

The environment which has made you the degenerate that you are, is the artificial, poisoned environment of civilization, in which conditions are such, says Carrel, "as to render life itself impossible." If you are not satisfied with your present state of physical, moral and mental degeneracy and desire to improve, you should remember that you cannot safely change yourself without changing the world in which you live.

Lesson No. 27
The Common Cold

"Science Given Fund to Find A Cold Cure, Baltimore, January 10th (A.P.) — A gift of $195,000.00 to John Hopkins University for the study of 'the origin and possible cure of common colds' was announced today. The gift, to be known as 'the John J. Abel Fund,' was made by the chemical foundation." — Daily press of January 11th, 1928.

Nearly a quarter of a century ago that announcement appeared in the daily press. Today knowledge as to "the origin and possible cure of common colds" is right where it was then. To show how little is known of the common cold, a certain doctor, writing on the subject, in Nature's Path in 1945, said: "According to the best treatise of orthodox therapy, medical science does not yet know what a cold is nor how it originates. It has been designated at one of Nature's great mysteries. The common cold is a provision of Nature devised to conserve our energy and vitality. In reality, it is one of Nature's most widely bestowed blessings. . . . The common cold aids the body."

This doctor, with many others, believes that the mucus expelled during a cold represents waste that has accumulated in the body. So he regards a cold as a purging process. This doctor would smile if told that the common cold is the first definite signal of the body's intelligence to warn the victim that polluted air is flowing through his nose into his lungs. He never heard that before. When this doctor was a little child, the common cold was the first warning signal that he had started down the well-traveled road of degeneration that leads to an early grave.

Degenerative Process

The degenerative process begins in the seed, but we will follow it from time of birth. To you a baby is born, and you are happy. You begin to plan its future and want to see it grow into a fine man or woman. You have not been taught that your home and your environment are saturated with polluted air, and that a flag containing Cross Bones and Skull should be flown in the center of every city to warn its inhabitants that they live and labor in a sea of poison.

This poisonous air begins to eat away at the infant's air organs as soon as born. Evidence of this appears in the first little cold, the first sneeze, the first cough, the first mucus flowing from the little nose—these are the first signs that poisonous air has begun its destructive work. You are not told that vital statistics show the chief cause of death in children up to the tenth year is ailments of the air organs. The poisoned air of home and environment works fast and fatally. The poisoned air begins immediately to eat away the lining membrane of the air organs, the nose, throat, tonsils, sinuses, trachea, bronchi and lungs. By the time the child is four or five years old, the damage has gone so far it can be revealed by X-ray examination, which shows white spots in the lungs. Some call these white spots "small stone coffins in which are buried the germs of tuberculosis."

Those white spots are actually the remains of ruptured air cells that have healed. But they have lost their function. No more can they function as air cells. That early in life they are done. Their function is "gone with the wind." To that extent the lung capacity has been decreased. To the same extent the vitality has been decreased. Your body begins to weaken and move toward the grave when you are a little child. In those "small stone coffins" are buried the worthless remains of precious air cells which did their allotted part in forming the connecting link with the Cosmic Source of all things, by drawing into the lungs and digesting the Divine Breath of Life and passing it, with its precious cargo of vitality, on to the blood, nerves and lungs, but which cells are now as useless as a paralyzed arm. In this age of smoking fathers and mothers, the children get regular doses of tobacco smoke, and what it does Dr. H. Bieler states as follows: "When tobacco smoke is inhaled, by children and others, the irritating process occurs, but is not quickly felt because the lungs do not have a sensitive network of nerves. The lungs become red and inflamed, but the condition must grow serious before it can be felt because of the deficiency of sensory nerves in the lungs. The absence of pain makes one oblivious to the damage being done. One is unaware that one's lung lymphatics are getting black with tarry irritants, and that the actual breathing capacity of the lung air-cells is soon diminished more than one half' (In Fact, July 1943). With the breath gone the life goes. Half of the body's vitality is gone when half of the lung capacity is lost.

Polluted Air

Medical literature contains no information of damage done to the lungs and body by poisoned air through the years. If germs are the cause of disease, poisoned air has nothing to do with the matter. The germ theory is protected by no study being made of the effect of polluted air on the body. To sustain life and preserve the body, the air must be the kind that man is made to breathe; and he is not made to breathe the filth, dirt, dust, smoke, soot and poisonous gases and acids of city air.

Professor H. Landsberg, Geophysical Laboratory, Pennsylvania State College, made a report on the Studies of Air Suspensions," in which he said: "Wherever human dwellings are, wherever industry has found a foot-hold, the numbers of dust particles in the air are vastly increased, and added to the list are many substances of high chemical activity. Among the more dangerous compounds in the air, nitric and sulphuric acids are always present in combustion gases. The constant irritation of the tissues of the respiratory organs is obvious when it is known that in city air about 900,000,000 of these nuclei pass through these organs daily, of which some 90,000,000 are retained in nose, throat, and lungs."

Ninety million enemies of health and life, inhaled daily out of city air and retained in the air organs are enough to wreck any organism in time and send it to the grave. The common cold is one of the symptoms of the damage being done. Yet the doctor told us that the common cold "is one of Nature's most widely bestowed blessings." A scientific study of city air made by the Temperature Research Foundation of the Kelvinator Corporation showed that: 1. The average dust-fall in a large city is approximately 230 tons per square mile per month. 2. An average of 2,500,000 particles per cubic inch are constantly present in city air. 3. More than 12 pounds of various dust particles are precipitated in the lungs of the average city dweller in a year.

Nashville, Tennessee, is not considered a smoky city. A press report stated that a survey of its air showed more than two tons of soot per square mile fell from January 15th, to February 15, 1938. Professor O. C. Gray, Engineer of the Bureau of City Tests at the University of Cincinnati, measured the filth fall in that city for the month of October 1930. In the business section the fall for the month mentioned totaled

1,176 tons per square mile. In the filth were 2.72 tons of tarry matter, 161 tons of carbon, and 1,012 tons of ash.

Gray said that city air is a deadly compound of smoke, soot, fine particles of dust, glass, rubber, mixed with acids and fumes, which included carbon monoxide, sulphuric acid, hydrochloric acid, hydrocyanic acid, nitric acid, benzene, methane, sulphur and other dangerous chemicals. Being thus informed, one can understand reports such as that in the press of December 14th, 1945, that "The sharpest weekly rise in influenza and pneumonia cases since the 1943 epidemic was reported today by the U.S.P.H.S."

The most dangerous disorders are those affecting the air organs. The death-rate in pneumonia cases is appalling. In the 1918-19 Flu-Pneumonia epidemic, 95 patients out of every 1,000 cases of influenza in New York City were lost and 640 out of every 1,000 cases of pneumonia. In the U. S. Army at that time 345 of those young, vigorous soldiers died in every 1,000 cases of pneumonia. (To Combat Disease is Dangerous, P. 35 by Dr. George R. Clements).

Dr. Arthur Vos, M.D., wrote: "Of all the food required by the body, 90 percent must be oxygen. A man weighing 150 pounds is composed of 110 pounds of oxygen by weight if the oxygen contained in his body were set free, it is estimated that it would fill 750 cubic inches of space." — Philosophy of Health. Dr. E. E. Marin, M.D., said: "Science tells us that over half of our maintenance comes in the air and oxygen we inhale. This means that pure air is much more important than the food we eat. We can live without food for three months without much inconvenience, but we can hardly live one minute without breathing" (Truth Teller, January 1940). Dr. Thomas Darlington, former health commissioner of New York City stated: "The products of combustion irritate the eyes, nose, throat, the respiratory tract, bronchial tubes, and gastro-intestinal area. In the lungs the carbon particles become imbedded in the air cells, and in time the lungs change in color from natural pink to black. I have performed many autopsies upon New Yorkers and almost without exception their lungs were as black as night. There is a striking parallel between smokiness of cities and higher pneumonia mortality. The soot, having coated the interior of the lungs, obstructs their natural eliminative processes and the flow of oxygen into the blood" (Quoted by W. B. Courtney in "Our Smoky Cities," in Collier's).

In 1945, 424,328 persons died of what doctors term heart disease. The cause was polluted air. In 1945, 174,640 persons died of what doctors term cancer. In this condition polluted air plays a leading part. In 1945, 129,144 persons, including the late President Roosevelt, died of what doctors term brain hemorrhage. This condition is largely the work of polluted air. The press of March 6th, 1944, reported that coal gas from the locomotive of a freight stalled in a tunnel in Italy killed 500 persons. Only 49 lived to be taken to hospital.

A case is reported that 146 British men and women taken prisoners and locked in a small room that had only one window. The opening was too small to supply fresh air for so many lungs, and by morning 123 were dead. Before these victims died they suffered from "shortness of breath," then became unconscious, and expired gasping for breath, just as you will do when you die. The press of October 26th, 1945, reported that in Los Angeles "heavy clouds of smoke clinging close to the ground, mixed with stinging fumes that caused people to gasp for breath, prevailed in Los Angeles this morning." The account said: "The manager of one firm called up the Sanitation Director and told him that his employees threatened to walk off the job because their eyes smarted and they had choking sensations in their throats because of the 'gas attack'."

In the November 1943 issue of his publication, the Editor of *Let's Live* wrote: "This (Los Angeles) area has heavy fogs that hold down the fumes, prevent eliminating breezes and obstruct the healthful ultra-violet and other rays. Hiking to the top of Mt. Hollywood, I have seen the black pall of smoke hanging over the city (Los Angeles). I even went aloft in airplanes and found this dense shroud of deadly smoke, dust and fumes make a ceiling 1,500 to 2,000 feet."

The hemoglobin of the blood has an affinity for carbon monoxide gas approximately 300 times greater than for oxygen. The air in the cities and on the highways where there is much traffic is so saturated with carbon monoxide gas, that the blood becomes only part oxygen-hemoglobin while the other part becomes carbon monoxide hemoglobin. This lack of oxygen makes people pale, weak, anemic, dizzy — and they look to food for relief.

Acute Ailments

Almost all acute ailments start with the common cold. The basic cause is polluted air. Practically without warning the nose will often begin to drip and one feels awful in just a few moments. Polluted air frequently works that fast. The patient is rushed to bed. The bedroom is often filled with polluted air and the bed is emitting poisonous fumes and odors. So a bad case of influenza or pneumonia may result.

Whole communities may come down with influenza and pneumonia in winter. The press of February 21st, 1943, reported that at Coshocton, Ohio, 1,500 persons were ill with influenza and all public schools in the city were closed. The press of March 31st, 1936, reported that in Milwaukee "one-fifth of the city's population was ill with influenza — 120,000 persons." The press of December 8th, 1943, reported that in Washington, D. C., 100,000 were ill with influenza; that Philadelphia had 200.000 cases, and Louisville, Kentucky had 25,000 cases. In Newark, New Jersey, 200 school teachers were ill with influenza, and in Detroit, 228 members of the police department had it. The press also said, "flu kills 2,000 in Berlin in a week." Nothing strange about this when the facts are known. We wonder why more are not sick.

The "common cold" is considered a very simple ailment because so little is known about the physiology of the body and its requirements, and nothing about the spiritual centers of man. As the delicate lining of the sinuses become inflamed, it becomes swollen and congested. To relieve the misery almost everything need from paralyzing drugs to poisonous sprays. That is the scientific way the spiritual function of the sinuses is destroyed, usually while man is still a child. Polluted air, ailments of the air organs, then come the poisonous remedies, and man in modern civilization is reduced to a purely physical entity because his spiritual centers are destroyed. He is then limited in his acquisition of knowledge to his physical senses, which usually deceive him.

Air Cells Burst

When polluted air is destroying the air cells of the lungs, causing congestion and pains in the chest, that is called a "Chest Cold" and remedies are offered for the condition while no one seems to know the cause of the condition. If the air is sufficiently foul, the lung cells grow inflamed and swollen, producing tightness in the chest. Some cells burst, and then their air function is forever gone. As the larger cells burst, the victim coughs up blood that seeps into the air cavity of the lungs.

Polluted air is the cause of nasal stuffiness, sinusitis, laryngitis, tonsillitis, diphtheria, bronchitis, mumps, hay fever, measles, scarlet fever, chickenpox, smallpox, influenza, pneumonia, and the chronic conditions of asthma, cancer, tuberculosis, etc.

Vital Adjustment

In Lesson 4 we discussed Vital Adjustment. The body, vigorous in youth, has a health standard at birth much above that of the environment. So an adjustment must occur to produce harmony. The state of the environment is fixed and stationary. It cannot be changed. Any change to produce harmony must be made by the body. So by the means of venous disorders of a degenerative character, the body begins its long, painful adjustment to the low health standard of the environment — where it is shocking to know that "civilization has created conditions of existence, declared Carrel, "which render life itself impossible" (P. 28).

The mechanism of this adaptative process begins to work in the body as soon as the child is born. By a long series of sickness, called children's diseases and beginning with the common cold, the vital body of youth is weakened and reduced in vitality to the harmonious level of the low health standard of the environment. Perfect correspondence between the body and its environment must prevail; and here is how that condition is produced. The adjustment does not come suddenly. It is a slow process, instituted by Infinite Intelligence to prolong the life of the body as much as possible. Were the body's constitution so rigid and unyielding that this process of vital reduction, of degeneration, could not occur, the shock of the health-destroying agencies of the environment would cause sudden death. It does occasionally, as where a baby dies in its crib from breathing polluted air in the home. The press of June 6th, 1945, reported that a young couple solicitously moved their baby's crib into the kitchen and lighted the gas range oven to keep the infant warm. Later they found the baby dead of carbon monoxide poisoning.

Each illness through one's life, from the common cold onward, is a step down the ladder of degeneration to the grave. If the illness is slight, the downward step is short. If severe, the downward step is longer.

No Complete Recovery

All recoveries from each illness are only partial — unless one makes a complete change in habits and environment. You change your world by changing your habits and environment. If you continue in the same path without change, you sink in degeneration. There is no return unless you act while there is yet time and make the change mentioned. Otherwise, you go down, down, down — and you never come up again. You never return to that point where you were prior to your first illness.

Comes the day when you may have influenza or pneumonia. They are the same, differing only in degree of intensity. Influenza is a bad cold, and pneumonia is a worse one. You may not die, but you will experience a weakness that will remain, if you have reached mid-life; and it will remind you that you are sinking. Some call it the work of the aging process. As your lung capacity decreases because of ruptured air cells, your vitality decreases in the same ratio. You see your energy fading. You begin to slow up, to get wobbly on your feet. Old age they say. Barring accidents, internal poisoning, and the poisonous remedies of today, you will never die until your lungs have degenerated to a point where they can no longer take in enough of the Breath of Life to supply the body's requirements. The cause of nasal stuffiness is polluted air, which irritates the nasal membrane, resulting in congestion, swelling, and a flow of mucus. The result is a decrease in the nasal passages. The condition gets worse from continued breathing of polluted air.

The nasal passages are closed by congestion and swelling of the lining membrane, making mouth breathing necessary. Continued breathing of foul air makes the condition still more serious, and it extends into the throat, causing sore throat and hoarseness. This is the cause of diphtheria in children, which is said to be caused by germs, and to prevent it poisonous serums are injected into the children. Extending down the air-tube, the trachea, the foul air irritates it and the smaller tubes (bronchi) which branch off and enter the lungs. When the irritation, congestion and swelling reach the lungs, the patient is in grave danger. The victim now forces breathing hard and painful. This is lung stuffiness, and results from the same cause as nasal stuffiness. The body temperature rises (lung fever), — influenza or pneumonia. A leading medical doctor wrote: "Pneumonia is always with us, has a frightful mortality, and its lesson is always one of humility to the medical profession. It is doubtful whether the death rate today from pneumonia is any less than it was a thousand years ago." The breathing organs are so badly crippled by polluted air in the pneumonia stage that they are unable to do their work properly.

The suffering body cries out for oxygen. The air cells and air tubes of the lungs are slowly being closed by irritation, congestion and swelling. The patient now breathes hard and fast through his mouth, and his heart-rate increases. The lungs are rushing the blood through the heart and all over the body, striving to save life by supplying the necessary oxygen. The next step is to rush the patient to a hospital, where he is put in the "oxygen tent". The air in the hospital is filled with foul fumes, drug odors, tobacco smoke, motor car fumes, and what not. The patient may die quickly and it is often said to be a "heart attack."

Mucus is excreted freely by the air organs during a cold. That is the body's only protective measure against the irritation caused by polluted air. The mucus is not composed of accumulated body waste, as many claim, but of the serum of the blood, elaborated into mucus by the membrana mucosa of the air organs and excreted by the mucus glands.

Hardened Mucus

During a cold on the lungs, you cough up much of the mucus, but not all of it. The mucus which remains in the lungs aids in the process of lung degeneration. It remains in the tiny cells and tubes of the lungs and goes through a process of hardening in time, clogging the cells and tubes and they lose their breathing function. No more can the life-sustaining oxygen flow into them. To that extent your lung capacity and vital capacity have been decreased. As a child you could run and romp all day without tiring. That indicated good lungs. At 30 some begin to slip. At 40 more begin to go down, and a little exertion makes them pull and pant. At 50 the shortness of breath becomes painfully noticeable. The lungs are going. Poisoned air is doing its destructive work.

Sir Jonathan Hutchinson (1828-1913) noted English physician, discovered that man's vitality comes from the air he breathes, not from the food he eats; and he termed the lung capacity the Vital Capacity (Greene, P. 284). The respiratory capacity continues to increase as a rule until about the 35th year at the rate of five cubic inches per year. Then it should remain stationary, but it does not. Polluted air continues its destructive work. From the 35th to the 65th year the respiratory capacity diminishes, as a rule, at the rate of about 1½ cubic inches per year. The rate of diminution depends upon the kind of labor one performs and the kind of air. It is less in the case of the farmer working out in the open air, and more in the case of the clerk working in a stuffy office filled with tobacco smoke. As a general rule the lung capacity of a man of 60 is about 30 cubic inches less than when he was 40. There should be no such decrease. With the decrease in lung capacity comes a decrease in vitality. The decrease denotes the damage done by polluted air.

The man who finds himself growing short-winded, should know that lung degeneration is the cause, and that polluted air is the cause of lung degeneration. But he thinks it is caused by the earth's turning on its axis. That has nothing directly to do with it. Cite some facts says the skeptic. The press of October 5th 1937, stated: "In the depression, the death-rate in Pittsburgh decreased. Since prosperity started to return, and factories began pouring out more smoke, the pneumonia deaths are increasing sharply. In the depression they were 91.8 per 100,000. Today they are 167.4. Before 1927 the rate was 200." There we have the cause of colds, influenza, pneumonia, etc. But some say it is the work of germs.

An Invisible Foe

The press of December 8th, 1944, said: "An unseen enemy with which they never had to cope in their native jungle, struck down scores of wild animals and birds today in a lower Manhattan (New York) menagerie. Gas from a defective heater killed 24 specimens and overcame a dozen more before a watchman discovers their plight."

"SMOKY CHICAGO"
"Oh Chicago," said the star,
"How I wonder what you are;
"Hidden there beneath your cloak
"Of whirling soot and filthy smoke."
—John M Cutcheon, Jr.

With that verse M'Cutcheon begins in the Chicago Tribune of May 5th, 1946, a story about "Chicago's Smoke Screen." He says that the recorded soot-fall per square mile per month in Chicago for 1945 was 67.7 tons.

The Aging Process

As you weaken and wrinkles appear in your face from your ailments, you imagine you are growing old. We have shown the cause of your ailments and the cause of your increasing weakness. The dropping water wears away the stone, not the revolution of the earth. Stop the action of the water and eons of Time may pass, but that will not affect the stone. It is not the imaginary flight of Time but the cumulative effects of your ailments that drag you down to decrepitude.

In the 'Precepts of Ptah-hotep." is the Papyrus Prisse said to be the oldest book in the world, a forbidding picture, of the miseries of old age are recorded from the mouth of Ptah-hotep himself when he was 110. He said: "The progress of decay changes into senility. Decay falls upon man and decline takes the place of youth. A vexation weighs upon him daily Sight fails; the ear becomes deaf; the strength disappears; the speech fails; the mind weakens, remembering not the day before. The whole body suffers. Taste disappears. Old age makes man altogether miserable. The nose is stopped, breathing no more from exhaustion."

In his word-picture this man does not say that sickness is responsible for the decline into senility. He terms it the progress of decay, and he is right. Decay of man's body results from definite causes, and the turning of the earth, on its axis is not one of them.

Notice the reference to the nose that is stopped and breathes no more from exhaustion. The nose is not exhausted. Poisoned air has caused the lining membrane to thicken, thus closing the air passages. The nose took all it could and lost its function because its air passages were closed by a thickened membrane caused by polluted air. You know not that you are constantly surrounded by an unseen enemy to health. You know not that you are constantly suffering from a mild case of blood poison.

The Blood

The condition of your body can be no better than that of your blood. You control the condition of your body by controlling that of your blood. Sickness is impossible if your blood and its circulation are normal. So health is impossible if your blood is poisoned by every breath you take. — poisoned by that invisible enemy of health by which you are constantly surrounded. Polluted air enters your body at every breath. It will cause every ailment that poisoned blood can produce, and that includes all ailments known. You can be happy only by keeping well, and you can easily keep well by keeping your blood normal.

You may not know that your blood is changing constantly. It changes completely three times in one day. By fasting one day, drinking nothing but clean rain water (or a fresh fruit juice diet), and breathing the best air found only in country regions, away from cities and highways. The blood is purified, and will immediately begin to build health. The law of HEALTH may be briefly summarized as follows: 1. Cosmic Rays, as an. are the creative agency and vitalizing force. 2. In health, the vitalizing force functions smoothly and insensibly. When its function is hindered, the body's intelligence increases its physiological powers in an effort to remove the obstructing object or condition. The super-active state thus produced in body function is falsely termed disease, and named according to the location of the most marked symptoms. There is good health and bad health, but no such thing as disease per se. 3. The body is created complete and perfect, wanting in nothing not supplied by the creative power, and incapable of receiving anything more from human hands. It is self-building, self-operating, self-regulating, self-preserving and self-repairing. 4. All the so-called healing power on earth is within the body itself. That power nothing can aid but the natural elements of the cosmos which are produced by the creative agencies. 5. The condition of the flesh depends upon the condition of the blood. Insofar as the blood is active and normal, and to that extent only, will and must the organs, tissues and cells remain healthy and function normally. 6. In exact ratio as the blood becomes stagnant and abnormal will and must all organs, tissues and cells show a decline from the normal state. That is what is called disease. 6. The healthful existence of the body depends upon the condition of the blood. Normal flow of normal blood maintains the body's healthful equilibrium. 8. Retarded circulation and polluted blood disrupts the body's healthful equilibrium. Remove the cause and the effect disappears. 9. Purification of the blood and acceleration of its movement is scientific treatment. There is no other. 10. The blood is readily purified and quickly normalized by the natural process of fasting, followed by every wild animal by instinct. No liquid should be taken but clean rain water or fresh fruit juice, and one should breathe the best air that can be found only in areas far removed from cities and highways.

Wonders Of The Air

Late discoveries in the field of atoms have scientists running to and fro as they begin to search out the wonders in the air. Cosmic rays and cosmic radiation are terms being used by the scientists as they delve into the wonders in the air. The late Sir James Jeans, F.R.S., was one of the first scientists to call attention to the "impact of cosmic rays upon human beings." He wrote: "Cosmic radiator falls on the earth in large quantities.... Every second it breaks up about twenty atoms in every cubic inch of our atmosphere and millions of atoms in our bodies every second, and as yet we do not know what its physiological effects are."

In 1935 Professor R. A. Millikan, one of America's foremost physicists, said: "Cosmic rays are raining enormously energetic bullets of some kind (Photons. electrons or both) from all directions upon the heads of mortals who live on the face of the earth." In 1939 Professor P.M.S. Blackett, F.R.S., stated that: "The earth is being bombarded by atomic particles of surprisingly high energy. . . We know almost nothing about the effects of cosmic radiation on man"

Professor Wilfred Branfield, in an article entitled "Continuous Creation," sought to show that the substance of living things comes from the air. As to Trees he said: "In tree life, so much comes from the air and so little from the soil.... Every changes every new intra-atomic spatial re-arrangements of protons, neutrons and whirling electrons, every addition or displacement of electrons, sets up vibratory resonance ... building up atoms of higher mass. . . The reactions are electrical, and it is useless and foolish to apply chemical methods."

The Prana of Yoga is the Cosmic Radiation of modern science, and the presence of radiant force is proven beyond disputation by instrumental indicators and recorders — above all by the use of the Geiger counter, the Compton cosmic-ray meter, sensitive electroscopes and specially prepared photographic emulsions. The highest powers of the body are Spiritual, and they fade out first in degeneration, caused by polluted air and dating from the birth of the infant. The nose, sinuses, trachea, bronchi and lungs are the Gas Chambers, the Life Organs. Cosmic Radiation, appearing as air, seems to be the Life Essence. For to stop the breath is to stop the life; when we cease to breathe we cease to live. If Life is God, then the breath is of God and in God. "With every breath we are linked with the Divine more closely than we realize." The mystery of Life itself, of all that is, may be discovered by studying the Breath of Life (Gen. 2:7).

In civilization the Breath of Life is polluted beyond description, and the Life Organs of civilized man are crippled by polluted air while he is a little child.

Lesson No. 28
Cosmic Air Purifiers

The press of August 19th, 1939, stated that gases and acids in the air of the Paris (France) area were: "Eating away and disintegrating the historic monuments of that city. The rapid decay of these stone monuments dates from about 1800, since then the smoke and fumes from factories, river tugs, motor cars and trucks and heating plants have steadily increased. The smoke, mixed with the exhaust of motor cars, trucks and buses, produces a compound of sulphuric acid gas that chemically attacks everything it strikes."

The air must constantly be purified; even in the country the air becomes foul. In desert regions, like New Mexico, Arizona and southern California, the air becomes ladened with dust particles and is bad to breathe. There are seven cosmic agencies of air purification, as follows: 1. Rain 2. Wind and hurricane 3. Vegetation 4. Earth rays 5. Violet rays 6. Cosmic rays 7. Electric radiation

Air is washed and cleansed by rain. Places that have much rain have much pure air. Places with little rain have less pure air, unless located in high regions or near the sea. In the midwest, and especially in the desert region of the south-west, with little rain and lots of desert dust, dirty air kills thousands. The press in 1935 reported that "70 persons died of 'dust pneumonia' in ten days in one small community, and asthma, tuberculosis and varying disorders of nose, throat and lungs are on the increase."

Winds and hurricanes are purifying forces. They send the stagnant air flying in streams that dissipate the impurities in it. On windy days the highways and cities are purged of their polluted air. The purification lasts only until the wind subsides. The press of October 6th, 1937, quotes Dr. Haythorn and Dr. Schnurer on this point, as follows: "High winds blow pneumonia out of cities on the wings of smoke in winter. Calm days, in smoky cities, are usually followed in about fifteen days by a rise in pneumonia."

These doctors prepared a chart of winds which showed a frequent rise in deaths from pneumonia fifteen days after calms in Pittsburgh, and a fall after good blows. There is the evidence to prove that the foul air in cities causes sickness and death. In the Carboniferous Age the atmosphere was highly charged with carbon dioxide gas. Ferns were abundant, some being as tall as trees. In the carboniferous forests grew the Lycopods or club-mosses, now represented by insignificant forms, but then growing sometimes 75 feet high or more, with trunks three feet in diameter, and spreading branches. The Vegetable Kingdom is not strictly an air purifier, but an air conditioner. Plants possess the power of absorbing carbon dioxide, assimilating the carbon and rejecting the oxygen in a free state, thus making the air suitable for the larger land animals.

Only after millions of years of plant life on earth, say science, was there enough oxygen in the atmosphere of the earth to support the higher forms of land animals. This fact constraints some authors to assert that forest regions are the most healthful areas for man. Thus, we see how man disturbs nature's equilibrium by destroying the forests to provide fields for his artificial crops. The purest air comes from the ocean where there is nothing to pollute it. The foulest air is found in cities. The larger the city the fouler the air. In the wide open spaces of country and sea, the other four agencies continuously create Ozone to clarify and purify the air. When man is not polluting the air with the fumes of his fires and inventions, they keep it in good condition. They also help destroy harmful gases and acids produced by man's work. As they were never intended to do so, they cannot overcome the excessive air pollution of homes, hospitals, factories, stores and city streets. These cosmic agencies cannot break through the layers of poisonous gases and acids, created by man's work, with sufficient force to convert oxygen into the essential quantities of Ozone.

Outside air in the country, or in fairly open spaces, such as a high-class suburban residential district, comparatively free of motor car and truck traffic, while it may seem pure, is still far from being the activated, ozonated air intended for man as the Breath of Life.

The air of the average home, in city or country: 1. Is saturated with carbon monoxide gas from fires, cigars, cigarettes, cookery fumes, fumes of motor cars, tracks, gas engines, locomotives, etc. 2. Is saturated with the fumes of filthy barnyards and hog-pens, farm tractors, and sprays used on gardens and groves. 3. Is saturated with the fumes of garbage and paint cans. 4. Is saturated with the fumes of refuse from the streets. 5. Is filled with tiny particles of filth flying in the air. 6. Is saturated with numerous gases and acids, with small particles of solids, wafted by the winds for miles in all directions.

The Home

Put up a roof and you have obstructed three of the natural ozone-creating forces. Add the walls, and you entrap the fumes flowing in from without and those generated within by man and his work. The hermit's hut in the hills would soon be filled with polluted air unless doors and windows are kept open all the time. Filth that destroys health and shortens life is constantly generated and eliminated by the body itself.

Your bed should be left open all day to dissipate the filth generated and eliminated by the body during the night. The sun's purifying rays should shine in the bed several hours each day, and pure air should fan it for hours before it is made up. It is much better of course to hang the bedding on a line in the air and sunshine. The ordinary bed is not fit to sleep in. The material of which beds are made, including the feathers in the pillows, go through a steady process of decay, and constantly emit foul, musty odors that are injurious to health, but not noticed because one gets used to them. Furthermore, the polluted air of civilization has ruined the sense of smell in most people. Insomnia can often be traced to foul air in the bedroom, some of which comes from the bed itself. That is the language of the body's intelligence as it tries to tell you to move into better air. But you believe in taking sleeping-tablets, and thus force your body to endure a condition that is slowly destroying it. Those with weak lungs do fairly well in the good air during the day, but have trouble and cough after they have been in bed long enough for the foul fumes of the room, bed and bedding to fill their lungs. Some have serious coughing spells during, the night and find it hard to breathe — all due to the damage done to their air organs by the foul fumes of their bed, bedding and bedroom.

This philosophy is too new to impress many, but one will soon be convinced if it is given a trial. Just as sure as you live and sleep in the fresh outside air, you will see health improvement.

Where To Live And Sleep

Man is an air animal and is constituted to live and sleep outdoors. Primitive man lived in the open air, and slept on some dry grass or leaves in the open under the trees. His bed was well aired very day.

The author in 1898-1901 while in the Philippines, emulated primitive man by living outdoors and sleeping under the trees. During all that time he never had a cough or a cold, and when he came home it was hard at first for him to sleep indoors. He had to get used to it again. This outdoor life in the Philippines was a valuable experience. He often had to sleep in wet clothing and thought he would die of pneumonia during the night as he had been taught such things. As he rose next morning feeling as fit as a fiddle, he saw how wrong these teachings were. He found from whence came sneezes, coughs, colds, sore throat, bronchitis, tonsillitis, hay fever, influenza, pneumonia, asthma, tuberculosis — all disorders of the air organs. He owes much to that lesson he learnt by living the outdoor life in Luzon, and has applied it through the ears. He swears by the outside, pure air, and now in his 74th year he enjoys the health and vitality which that air has given him.

The press of March 2nd, 1943, said: "Ensign P. C. Nolan, commanding a gun crew on a merchant vessel, spent 39 days at sea in an open boat after the vessel was torpedoed. Every minute of the 39 days he was soaking wet and badly chilled, but he reached a South American port in good health." One test case is sufficient to show that it is an error to put the cause of certain ailments on the weather. Wild animals live out in the cold winter rains and snows, sleep in the snow, and no doubt become badly chilled; but no hunter ever saw me of these animals suffering from ailments of their air organs. You cannot breathe filthy air and have good health. The environment, which includes the atmosphere, must be clean and pure.

Where shall I live to have good health? In this civilization that becomes the biggest of all problems. In the words of the great Carrel, "Our civilization . . . has created certain conditions of existence which ... render life itself impossible" (*Man The Unknown,* P. 28)

Ozone

For several years, certain scientists have been delving into the secret of a cosmic gas ailed Ozone, an allotropic form of oxygen. There is a paucity of information on the abject. Some think Ozone is a filter of the sun's rays as they pass to the earth. Its molecules are believed to contain three or more atoms of oxygen. For that reason, it is referred to as 0/3.

Ozone may be prepared by passing a charge of electricity through molecular oxygen, forming a polyvalent. unstable form of oxygen. It is a colorless gas with a peculiar odor, from which its name is derived — Ozone, meaning "to smell." It is claimed to be 1/5th as heavy as oxygen, yet it resembles ordinary oxygen in its chemical content, although it registers a greater degree of activity.

Being in its polyvalent, unstable form, Ozone tends to disintegrate rapidly into molecular oxygen. If uncombined, the free atomic oxygen thus liberated is highly effective in promoting oxidation. In this sense, Ozone has several uses: 1. As a bleaching agent. 2. As a means of purifying water. 3. As an extremely active oxidizing agent. 4. As a powerful disinfectant and germicide. 5. As an effective agent in dissolving various abnormal deposits, as in cases of arthritis, nephorolithiasis (kidney gravel), and cholethiasis (gravel in biliary ducts).

Oxygen unites with the iron faction of the hemoglobin, loosely forming oxyhemoglobin. Ozone has been found to increase this process, which means more of what the body needs. Cell activity determines the amount of oxygen required. But the amount of oxygen supplied does not determine the activity of the cell. Cell function must always continue, and can continue in such low oxygen content that even a match will not burn. But if a process be abnormal, more oxygen is needed. If the oxygen is not supplied, the condition is termed anoxia (deficient aerator of the blood). It can occur as a result of improper external respiration, or internal respiration, or transportation by the body fluids. An acute inflammatory condition results in a demand for more oxygen. The body intelligence tries to compensate for the demand by increasing the respiratory rate.

Lack of oxygen results in a degenerative process that leads to calcification (hardening). Ozone, in its unstable, polyvalent form, will readily break up and form a more stable molecular oxygen. For this reason, in instances where there is a process requiring more oxygen, ozone is the answer. This was shown by the use of one of the largest ozonating systems in the Central London Railway. It was reported that during a severe influenza epidemic, the motor drivers who run through the tubes daily were free of the disorder. Clinical evidence has shown favorable results in the use of Ozone in both acute and chronic inflammatory conditions. It has been known to dissipate certain calcifications in arthritis.

In the case of a chronic condition as arthritis, if additional oxygen is introduced into the abnormal, oxygen depleted tissue, the process would be a reversal, and result in the removal of the abnormal deposits. In a small way, Ozone is being used in hospitals and in sterilizing water systems. But its use is still in the embryonic stage of development. Should the development of the use of natural agencies be successful it would interfere with the use of drugs, vaccines and serums. Then there would be the dilemma of presenting this drugless system and showing that the use of drugs, vaccines and serums are not only ineffective as remedies, but actually injurious. As Ozone is only 1/5 as heavy as oxygen, it rises and the air of the higher regions is more heavily charged with it. Knowing this secret of Nature, the Ancient Masters dwelt in the ozonated air of high places. Good health and long life were the result.

Ionized Air

It required radio-radar-television and the atomic bomb to get modern science interested in the wonders of the air we breathe. Previous to these inventions no one had believed there were mysteries in the air which we have now discovered. The press on December 7th, 1938, stated that Dr. F. Behounek and Dr. J. Klestchka, two scientists of the University of Prague, issued a report disclosing some of the secrets of the atomic gases of the air. They said that "the mountain climate is characterized always by a greater ionization than that of the lowlands." By "ionization" is meant the presence in the air of "ions", which are said to be electrified particles.

It appears that the atoms of the air gases, such as nitrogen, oxygen, ozone, carbon dioxide, are not electrical when they are whole. When ultra-violet rays, cosmic rays, radium, X-rays, and fast electrons smash the atoms, the broken bits become "ions," or electrified particles. According to Sir James Jeans, F.R.S., the air one inhales contains over 34,000 atoms per cubic foot that are broken up by cosmic rays, and thus become electrified particles. "This is the source of Vital Force," says one authority on the subject. In a recent letter to us, the writer says: "It is remarkable how many tribes and groups of people on the southern slope of the Himalaya mountains have been reported as extremely healthy and long lived.

I wonder whether the elevation, plus the protection the high mountain must give against sudden severe cold spells from the north, plus a moderate climate as to temperature, don't combine to prolong life there. I think we need a good book about those tribes, telling how they live, about their food, climate, habits, etc." Propaganda has made most health seekers so food conscious that they never think of air. They never ask whether the quality of the air has any effect on the body. All mysteries fade out when the simple facts are known.

Scientists who have investigated the air of various regions, high and low, find "the mountain climate is characterized always by a greater ionization than that of the lowlands," as above stated. People so fortunate as to live in the higher regions and breathe the better air, highly charged with ozone and electrified ions, experience an electric rechargement of the vitalization of their body cells. As this increases, in the same ratio decreases the desire and need for food, and one's *health improves.*

Shallow And Deep Breathing

A certain work titled "The Prana of Yoga" states that Prana is not merely cosmic rays of the atmosphere, but also ionized minerals that come from cosmic sources. As we breathe the air we inhale these ionized minerals, as well as nitrogen, which is transformed into protein in the body. The protein we eat never becomes the protein of the body.

As previously stated, Behounek and Kletschka contended that "the mountain climate is characterized always by a greater ionization than that of the lowlands." When man changes from his low environment and moves to a higher one, say to 5,000 feet, he finds that at first he breathes deeper and harder. Those knowing so little of the function of respiration, would advise him to return to a lower level before he dropped dead of heart attack.

The body requires just so much air for its needs. The amount is supplied by shallow breathing in the dense air of low altitudes. Consequently, those born and reared in the low levels are made shallow breathers by the dense air of their environment. As a result, their lungs are never fully developed. Millions of air cells in their lungs remain dormant and inactive. As such persons move into a higher region, they begin to breathe deeper and harder, with an increase in heart rate and a quickening of all body functions. This is the first effect, the temporary disturbance as the body begins to adjust itself to the thinner air and lower atmospheric pressure of high levels. A change in environment always causes a corresponding change in the body's function. A slight change in environment causes a slight change in body function that is not noticeable at first, but shows up in time. It requires more of the thinner air of high regions to supply the body's needs. The result is that function changes as the body adapts itself to the new condition. At the same time the body also receives more of the ionized minerals that come from cosmic sources. The effect of this is to lessen and dull the appetite for physical food.

From a shallow breather at low levels one naturally becomes a deeper breather at high levels. Millions of dormant air cells in the deep regions of the lungs are resurrected from their dormancy and become active, as they should have been from the first. This is another case of vital adjustment, but this time it is for the better. The lever is reversed, and the process of regeneration begins, being the secondary effect experienced by the body as man moves to higher elevations. Few understand this secret of physiology. Carrel briefly noted body changes, but this is one that even he failed to find. The general ignorance of this physiological secret of respiration is the reason people are advised to be careful and not exert themselves at high levels, and to return quickly to lower elevations when they notice these symptoms. These symptoms are really the signs of the regenerative process going into action. Remain at the higher elevation until the body has time to adapt itself to the thinner air, and no disturbance in breathing will be felt. The lungs will gradually expand to meet the new condition, and improved health will result, provided that all the other rules of health are observed. Here we find another secret of Nature.

When man moves to higher elevations and breathes deeper and inhales more of the ionized minerals that come from cosmic sources, he will soon notice his desire for food waning as the regenerative adjustment proceeds. He eats less as he breathes deeper of the thinner air, and absorbs more of the ionized minerals in the atmosphere.

Usually it is considered as bad any waning, weakening of the appetite, by those who do not understand, and they are urged to take a tonic or something to whip up the appetite. This reversal of the lever that changes degeneration to regeneration produces changes in the body that few can understand. Conditions of regeneration are so rare that those who have no knowledge of accompanying symptoms, usually consider the symptoms as bad and do everything they can to eliminate them. The symptoms cannot be suppressed without stopping the regenerative process. Many persons who experience these symptoms become frightened and get back to their accustomed lower levels. We experience discomfort as the body changes under the law of degeneration. Hence we must experience discomfort as the body changes under the law of regeneration, and begins to rise to a higher level of health. The fasting man experiences much discomfort as his body purges itself of clogging waste and improves its condition.

Not understanding this secret of physiology, the fasting man is advised, by those who should know better, to eat or he will die. We have a report in the case of a certain man who made the change from a low to a much higher altitude, and this is what he said: "The longer I remained in the higher altitude, the less food I wanted, and food that had been delicious now became disgusting. My body, receiving more cosmic food from the atmosphere, had less need of the gross physical food of the earth. Then I changed back to a lower level, where the atmosphere is deficient in cosmic food, as a result of which a ravenous appetite appeared, as my body called for more physical food from earthly sources, in a secondhand form, to replace what I had formerly secured from cosmic minerals of the air at the higher level. Toward the latter part of my stay in the higher altitude, I lost my appetite for food to such extent, that I am convinced from my experience that Breatharianism in the higher regions must have been the original state of man."

Breathe More — Eat Less

The answer to the above questions appears in the purer and better air of the remote, higher regions, and in the eating of less food – contrary to the theories of those who eat freely and often to keep up their strength and die as a result. According to ancient arcane science, one of the highly beneficial effects of sustaining the body completely on the chemical elements of Cosmic Radiation and the consumption of no food, is the development of the higher powers that are now dormant in the man who eats ... *his marvelous unused powers.* The Ancient Masters taught that food in the alimentary tract interferes with the natural use by the body of the chemical elements of Cosmic Radiation. Food in the body insulates the body against the natural contact of Cosmic Radiation by corroding the poles of the cells. That obstruction causes the function of the body cells to decline below the higher level. As this damaging condition of insulation is increased by much food and regular eating, Vital Force gradually decreases and decrepitude slowly appears. Then comes that time in due course when the function of the body cells declines more and falls below the Life level of vibration, resulting in the condition called death.

This explains the secret and little known reason why man's health and all his powers increase and grow more acute under a fast. During a fast, the alimentary tract becomes free of the insulating effect of food, and the cells are able to free themselves of the damaging corrosion. This permits the body to make more and better use of the chemical elements of Cosmic Radiation, produces an increase in all vibration, and puts the condition of the body back nearer to the normal state in which it was before man fell to the level of the animal plane by forming the habit of eating.

Air Is Life

Here is a true story which shows that Air is the Power of Life. A soldier on the western European front in World War II had a small cut in his throat caused by a sharp piece of steel from an exploded bomb. He fell in a heap and his buddy hurried to his aid. An examination showed no injury but a slit cut in the windpipe just below the larynx, causing the windpipe to close so air could not enter the lungs. Quickly the other soldier slipped his fountain pen into the trachea (windpipe) to hold it open so air could enter the victim's lungs. The wounded man immediately came to life, rose to his feet and walked as though nothing had happened. Walking jarred the pen out and it fell to the ground, and so did the soldier. He could not move and appeared dead. Again the pen was quickly inserted into the trachea, and again the dead man came back to life, rose to his feet and walked. This time the wounded man held the pen in place until he reached a first aid station, when a surgeon sewed up the injury, which soon healed and the soldier's life was saved.

The Living Stream that turns the Wheels of Life depends upon contrary, he knows that tobacco is poisonous and destroys him by the Breath of life. The condition of the body depends upon the blood, and the condition of the blood depends upon the Breath of Life. Pollute the blood and we plunge the body into degeneration. Pollute it more, and the body's function becomes an appalling convulsion which in due time ends in death. You may be pale, weak, anemic; you may be suffering from some terrible ailment, but when you go on a fast and breathe pure air, your blood will soon become normal and all parts of your body will soon exhibit renewed life.

Eternal Physical Life

Profesor J. S. Haldane, noted English astronomer, believed that Eternal Physical Life is possible. He wrote: "In years to come, when man learns how to live, he will never know illness, and will live for thousands of years. We now sit as the men in Plato's cave, with our backs to the light, seeing only shadows on the walls before us. Reality we never see. Living is actually a struggle for fresh air. Keep the vast lung surface of the organism supplied with fresh, unpolluted air, and also observe all other health rules, and there is no reason known to science why you should ever die. No matter how long you live, when you die your body will be young, and you will die for lack of oxygen. You will die because your blood cannot carry the required amount of Life-Sustaining Oxygen to the billions of cells in your body."

The New Age

Millions of dollars are spent to study cancer and other disorders but no doctor is ever employed to go out and study the cases of people who live long and publish his true findings. It is just as though the study of a stone would reveal the cause that pulls it to the earth when cast into the air. If we would learn how to live 150 and 200 years, we should study the lives of those who do it. But there is no profit health and longevity for those who profit from sickness. When some doctor is interested for himself, and at his own expense makes a study of these cases and publishes his true findings, he is silenced and liquidated and his report is discredited and destroyed as history shows.

Regardless of which road we take in matters of Health and Longevity, we find they all lead back to the Breath of Life. When man first began to eat, he knew that food was not needed to sustain his body.

He ate for pleasure and not from necessity. The same state is presented today in the case of the smoker. He knows that tobacco plays no part in sustaining his body. On the contrary, he knows that tobacco is poisonous and destroys him by degrees. But the habit of smoking is too strong for him to conquer. As long ages passed, the time came when man believed he had to eat to live. The eating habit had him in its grip, as the smoking habit has the smoker. His body had gradually adjusted itself to the practice, and craved food, as the smoker's body craves tobacco.

Old Age is man's oldest enemy. The Fountain of Youth is man's fondest dream. We have seen that senility is progressive degeneration. It is neither natural nor necessary. All its symptoms are pathological. The complex of these symptoms is also pathological. There should be a way to avoid it. We have seen that Fasting retards the speed of the aging process with a corresponding prolongation of the life-span. Why does man continue to look to food for the goal of his search, as he sees the body grow younger when no food is eaten?

For thousands of years the eating habit has held man in its grip. Current reports show that only a few in remote regions of the earth have escaped it. Carrel was great, but he could not rise above the influence of his medical training. He demonstrated that the body cells are immortal, yet he said they died. He should have known that immortal cells are above the nutrition level, yet he said they had to be fed. These errors he would have seen had his mind been clear and not clouded by false theories.

Immortal body cells neither eat nor die. They are the same category as the cosmic stars and planet. Body cells are composed of tiny stars that are as eternal and self-sustaining as the stars of the Cosmos For the cells are composed of similar electrons, atoms and molecules. It is the attempt to nourish the body that destroys it. It is the attempt to cure disease that destroys physical man. In its present state, the body depends upon the chemical stimulation of food and drink. If that state were natural, the condition of the body would not improve under a fast. That state is the result of ages of eating and drinking. It is an unnatural state resulting from the body's adjustment to a practice forced upon it. It was a case of meeting the practice by proper adjustment, or of perishing. The adjustment was made and resulted from a reduction of the body's vitality and a decline in its integrity and duration. This is devolution, degeneration, a process of decay, and somatic death is the end. Advanced students assert that man has reached the bottom of the downward trend.

He can go no lower. Hence the dawn of the approaching Golden Age is drawing near.

"Read and watch your world grow."
— George Bernard Shaw —

FIRST GREAT LAW
"The living body must function always in the direction of health, and sickness is the effect of obstructing natural function."

SECOND GREAT LAW
"Sickness is the result of the body's struggle to eliminate internal poisons resulting from bad environment and bad habits."

"Any good therefore, that
I can do,
Or any kindness that
I can show
To any human being
Let me do it now. Let me
Not defer it nor neglect it for
I shall not pass this way again."
Author Unknown —

"We squander health in search
of wealth;
We scheme and toil and save;
Then squander wealth in search of health
And all we get is a grave.
We live and boast of what
We own;
We die and only get a stone."

"MONEY WILL BUY.....
A bed but not asleep.
Books but not brains
Food but not an appetite.
Finery but not beauty.
A house but not a home.

> Medicine but not health
> Luxuries by not culture
> Amusements but not happiness
> A crucifix but not a Savior
> A church but not heaven."

"Though an inheritance of acres may be bequeathed, an inheritance of health and happiness cannot. The wealthy man may pay others for doing his work for him, but it is impossible to get or purchase health and happiness as many have found to their sorrow."

"The Laws under which we live are designed for our advantage. These laws are immutable and we cannot escape from their operation. It is in our power to place ourselves in harmony with them and thus express a life of health and happiness."

Lesson No. 29
Degeneracy Of Civilized Man

One of the greatest scientists of modern times wrote a book which he titled "Man The Unknown." In that book this great scientist asked this question: "How can we prevent the degeneracy of man in modern civilization? (P. 5). Carrel asked the question in deep seriousness, then answered it on Page 28, but apparently knew it not. He said: "Our civilization... has created conditions of existence which... render life itself impossible." If what we are pleased to call "civilization" has created conditions of existence which render life itself impossible, that explains why "man in modern civilization" is degenerating. Fifty years before this great scientist asked that question, Dr. S. A. Strahan, in an address before the British Association for the Advancement of Science, told why man in modern civilization is degenerating. He said: "All the deteriorating influences of modern civilized life tend toward the reduction of vital energy, and to the degeneration of the race" (Densmore, P. 380).

For half a century it has been known why man in modern civilization is degenerating, and yet nothing has been done about it by the health boards of the various states which spend the tax money of the people "to promote the public health." The artificial, degenerating, health-destroying, life-shortening conditions of living called civilization, are the direct cause of the degeneracy of man, including the poverty and misery by which he is surrounded, and nothing is done about it. Here may be the reason why: "The press of February 26th, 1948, stated that 'sickness brings the physicians of the U.S.A. $1,500,000.00 daily, according to a report of the director of the Public Health Nursing Service of the American Red Cross.' "According to an estimate broadcasted by the California Tuberculosis Association in 1934, sickness costs the people of the U.S.A. $15,729,925,398.00 a year — nearly sixteen billion dollars." No one can believe the health boards of the nation would dare to do anything to interfere with that golden stream of revenue.

Man's body is composed of condensed Cosmic Rays in the form of cells, and it is animated by cosmic ray in the form of gases called air. The lungs are the gas chambers, the Life Organs; the Cosmic Rays, as gases in the atmosphere are the Essence of Life. To stop the breath is to stop the Life. When we cease to breathe we cease to live. The internal lining of the lungs of those who dwell in the cities of civilization is black as coal (anthraconecrosis) with a coating of tar and carbon that seriously obstruct the free flow into the blood of the Vitalizing Gases of the air, and the body sinks into degeneration as a result. The logical consequences are that this man in modern civilization is steadily sinking, decaying, dying by slow degrees. If he drops dead, as many do, the doctors term it "a heart attack" and brush the incident aside as of small importance. Here is how the poisoned air of civilization knocks them out: 1. J.P. Morgan, New York Banker, age 75, died suddenly of a "heart attack." 2. J.P. Morgan's partner, only 45, dropped dead of a "heart attack." 3. Alfred McCann, noted food scientist, dropped dead of a "heart attack" in the bathroom of his home, age 52. 4. O. O. McIntyre, noted newspaper columnist, age 53, dropped dead of a heart attack." 5. Allen R. DeFoe, M. D., who became famous as the Dionne quintuplet's doctor, died of pneumonia in a hospital five minutes after admitted. 6. Edward Star Judd, M.D., chief of the surgical staff of the Mayo clinic and former president of the A.M.A died suddenly of pneumonia, age 57. 7. Dr. F. N. Wilson, University of Michigan heart specialist, died suddenly of a "heart attack." 8. John B. Swan, age 64, Soo Line engineer, died suddenly March 20th, 1994, of a "heart attack." 9. The press of June 4th, 1944, reported that Dr. C. E. Ryan, M.D, age 69, died of heart attack as he was delivering a baby.

To normalize and purify the vitiated, devitalizing atmosphere of civilization would require the destruction of civilization, of what constitutes civilization, such as filthy cities, the stinking barnyards and foul hog pens, the coal mines and other mines, oil wells, oil refineries, industrial plants, tobacco factories, gas engines, motor cars, trucks, buses, etc. That is the civilization which has created the conditions of existence which render life itself impossible (Carrel).

It appears from the findings and conclusion of Dr. Carrel, there is little worth saving in a civilization that has created conditions of existence which render life itself impossible. One or the other must go. Shall it be human life or civilization? It appears that a system of illusion, erroneously termed modern science, has badly blinded most people. Maybe we might learn something by reading what Carrel said about science in his 'Man The Unknown." This is what he wrote: "Science follows no plan. It develops at random. . . . It is not at all actuated by a desire to improve the state of human beings. ... Men of science know not where they are going. They are guided by chance" (P. 23).

When man left his native home in the hills and forests, where God originally put him, and began to build a false science on illusion and the artificial centers called cities, which constitute the heart of that condition termed civilization which produces poverty, want, sickness, despair, degeneration, slow death by degrees, war and slaughter, he set in motion the terrible monster that is slowly destroying him. The fury of the atomic bomb, unleashed in 1945, did more than destroy two Japanese cities; it destroyed the smug self-satisfaction that covered the world to reveal with what horrid brutality "civilization" is maintained.

Civilization

Wage earners are Economic Slaves, the lowest of which is the common laborer. We would not be far wrong to say that civilization is commercialism, and the latter is composed chiefly of capitalists and their hirelings. The hirelings are the Economic Slaves who are dependent on others for a living. They have no to time acquire knowledge, and are not noted for their wisdom. The position they fill is proof of their mental inferiority, and they are in the great majority in civilization. This vast majority constitutes the servants of Commercialism. They spend their days in weakening work and degrading toil. They have little opportunity for physical, mental and spiritual development. They must look for light to those who live on the fruits of their labor. If they found the true light it would soon end their days of economic slavery, so their mind, their education and religion must be controlled in order to keep them in darkness.

"The word 'Education' is a misnomer. The schools do not educate; they domesticate. People are trained like dogs and horses. They are trained to do the things and think the thoughts that those in authority want them to do and to think. They are trained to accept without prejudice and to obey without question the dictates of authority" (Humanity).

In the Golden Age publication of 1993 the editor of it reported that there were 150,000 homeless girls and women in the U.S.A. — female tramps in the greatest nation on earth. They were looking for work that they might labor to live. Economic slaves reduced to beggars in a Nation said by politicians to have the highest living standard in the world. That is just one of the sad consequences of modern civilization.

If we happen to have enough intelligence left, in this day of degeneracy, to recognize these facts and the fortitude to face them, we realize that the condition called civilization is a controlled, regimented system of living that is designed to darken the mind and control the man. It is constituted of an artificial environment from which everything natural is eliminated and populated by people so completely misled that they are incompetent to recognize these facts. Only the blind, dumb and deceived would join in the criticism of those who are still able to see these things in their true light, and have the courage to raise their voice in protest against them. When Carrel asked, "how can we prevent the degeneracy of man in modern civilization?" he had his answer in his own statement as follows: "The environment which molded the body and soul of our ancestors during many millenniums has now been replaced by another. This silent revolution has occurred almost without our noticing it. We have not realized its importance. Yet it is one of the most dramatic events in human history. For any modification in their surroundings inevitably and profoundly disturbs all living beings" (P. 10).

Carrel refers briefly to this new environment that replaced the old. He says it consists of cities with "monstrous edifices and of dark narrow streets full of gasoline fumes, coal dust, and toxic gases, torn by the noise of taxicabs, trucks, and trolleys, and thronged ceaselessly by great crowds. Obviously it (the city) was not planned for the good of its inhabitants" (P. 25). We can prevent the degeneracy of man in modern civilization only by removing the cause, and that cause is no mystery. Carrel has revealed part of it.

Alexis Carrel

In our work we quote often from the valuable book of Alexis Carrel, titled *Man The Unknown*, 16th Edition, Copyright 1935.

Carrel and Lakhovsky were the greatest scientists of the 20th century in matters of Life and Living; yet they were biased by medical training and occasionally went off at a tangent. Carrel studied the body, while Lakhovsky studied its animating force. The former showed that the body cells are immortal, and the latter showed that vitality is the action of the Breath of Life on the positive and negative poles of the cells.

Lakhovsky showed that as the poles of the cells become seriously corroded, they fail to respond to the electro-magnetic Breath of Life, and the cells fall in function below the Life Level of Vibration. That state is termed death. He did not put it in those words, but that is what he meant. When an attempt was made in the U.S.A. to translate into English Lakhovsky's great work, "The Secret of Life," the medical trust opposed it. When it was finally so translated, such a small number of the English edition was printed, that it was hard to get. We managed to get a photostatic copy from the Library of Congress, at considerable cost.

In his work, Carrel was amazed to find that so little is known of the body and its functions. He showed that almost all the theories in medical works as to Life and Living are erroneous, and he declared, "The childish, physio-chemical conceptions of human beings," advocated by medical art, "have to be definitely abandoned" (P. 108). He discovered that so little is actually known about man, that he titled his book "Man The Unknown."

Carrels exceptional skill was duly noted, and he was invited to become a staff member of the Rockefeller Institute of Medical Research in New York, and served as such from 1936 to 1939. In 1912 he won the Nobel Prize for his success in suturing blood vessels and transplanting organs in the body. His book we have mentioned is remarkable for two principal reasons: In showing that he was an extraordinary man, and in exposing and exploding the many ridiculous medical claim and preposterous medical theories, such as "Life is the expression of a strict of chemical changes" (Osler). His book should be read by all who still have faith in medicine and medical doctors. The work contains so much naked truth in exposing medical ignorance, that it made him unpopular with his medical brethren. That may have been the main cause he left the Rockefeller Institute in 1939, and the reason he was arrested in France in 1944 on a charge that he, as director of the Vichy-supported Carrel Institute of Paris, "Founded the organization for the purpose of supplanting the great French Universities, and of introducing Fascism and Marxism to students"-as the press of August 31st, 1944, reported under a Paris date line. It would appear more reasonable that Carrel was invited to instruct students in the real facts of Life and Living as discovered by him, but not taught in medical institutions. Carrel's sudden death in his 72nd year of "a heart attack," as reported in the press of November 6th, 1944, was the end of a great man and a great doctor. He was greater than his profession. His demise undoubtedly made the medical trust happy; but his book will live to plague the medical institutions, and to proclaim to the world the terrible truth that the system known as medical science is actually a blighting and controlled system of medical ignorance.

Carrel and Lakhovsky were both medical doctors, but the American Medical Association opposed them because their work exposed medical ignorance. Lakhovsky came to New York some years ago, and the A.M.A. gave him a cold shoulder. It was reported that he entered a hospital for treatment of a minor ailment, and was never seen again.

Lesson No. 30
Mysteries Of Life

"To the questions whence, whither, and to what purpose are we here, science gives us today an answer as empty as it did two thousand years ago." — Professor Harnack, University of Berlin.

Science is doing this and science is doing that. The facts show that science is running in circles and chasing phantoms. Ouspensky, a leading scientist, wrote a book of nearly 500 pages, titled "A New Model of the Universe." He exploded practically every theory of Modern science, and then capped the climax by declaring: "In reality, of course, no one knows anything" (P. 408). This scientist showed that we live in a world of illusion. That things are usually the reverse of what we think they are. Our senses deceive us. The sun seems to rise and set. The earth seems to stand still. Iron seems to be solid. Drugs seem to "cure" pain. They deaden the aching nerve while the cause of the pain remains.

Scientists make matters worse by limiting their investigation to things material. They refuse to believe in an invisible world, and grope in self-imposed darkness. They are materialists of the first order, and their theories of materialism are constantly exploding. Yet they refuse to believe in a spiritual world because they have condemned it as a fancy of ancient superstition. The result is that the word Science is losing the respect of students. It is coming to be more loosely used by schemers and more lightly regarded by scholars. The whole theory of materialism has been exploded, and scientists are busy trying to conceal the wreckage from public view. Science is defined in the dictionary as "accumulated and accepted knowledge that has been systematized and formulated with reference to the discovery of general truths and the operation of general laws." That definition does not describe modern science.

Modern science is not constituted of the "discovery of general truths and the operation of general laws," but of manmade formulations and rules, theories and hypotheses, opinions and speculations. The word Science does not mean the presentation of something new. It means the fund of knowledge and understanding of truth and laws that describe Cosmic Processes in their operations and productions. Science is positive knowledge based on truth and law. Yet little in the Universe is absolutely known. Man is ruled by his five physical senses. He can see and hear only within a very restricted range. He must depend for knowledge upon these deficient faculties, which are more limited now than ever before, as modern man is seriously degenerated both mentally and physically; and his technical instruments can never perceive Absolute Reality.

Modern Science Is Young

Modern science is in its infancy. It is just emerging from the gloom of the Dark Ages, when men were burned for possessing more knowledge than theology allowed. Some say it is six centuries old. Three centuries come closer to the mark. It was born under exceedingly adverse circumstances. Its coming was bitterly opposed by theology because its findings were feared. The pioneers of Modern Science, who dared to disclose discoveries that exploded theological dogmas, were quickly liquidated. They were tortured, assassinated, burned. Thinking men refused to swallow the theological dogmas of creation, supernaturalism, saviors, vicarious atonements, physical resurrections, purgatory, judgment day, and future punishment. Such dogmas were contrary to common sense, reason, logic, and the known laws of the Universe and these men were inspired to show that they are obviously false.

Theology represents the gospel Jesus as a supernatural being. That offends reason and creates heretics. It antagonized the thinkers who declared, 'There is no supernatural." And the fight was on. It is the use of the word "supernatural," and the preposterous propositions involved, which drive the rational mind to the other extreme. It is this that goads the scientific skeptic into characterizing as mere superstition, the loftiest perceptions of the Soul. The Agnostic is the product of theological dogma. So Modern science was driven into the position of being a movement to oppose the dogmas of theology; it obviously became atheistic and rejected the theory of a Cosmic Principle of Being, Cosmic Intelligence, Cosmic Law and Order. Its major premise came to be: 1. Life is the sum of body functions. 2. All is physical matter and mechanical energy. 3. The world is composed of blind and unknown forces. In opposing theology by assuming that everything has a physical basis, physical science has turned, twisted and distorted every fact of psychical phenomena and every fact of physical Nature. It has reduced man to a mere physical entity, and attempted to find physical causes of the "blind and unknown forces," which include what physical science classifies as the higher phenomena of life. So long as it seeks only to relate physical causes with physical effects, physical science is admirable. When it attempts to relate psychical phenomena with physical causes, it ignominiously fails. The inevitable logic of such assumption is to relate intelligence to the digestive organs, to depict Life as the expression of a series of chemical changes, and to define Love as the efflorescence of physical lust. Against such assumptions common intuition, common experience and common sense rebel.

The best intelligence of today accepts generally the physical facts of Nature as collated and classified by physical science. This same intelligence declines to accept the theories of physical science, as to the causes of psychical phenomena. These hypothetical dogmas, based upon only half the facts of Nature, confused even so great a mind as Huxley's. Having accepted the physical facts of Darwinism, he logically felt bound to accept Darwin's theories as to the causes of those facts. As a result, Huxley repudiated Nature and denounced it as a monster without a single principle that conserved justice, or love, or altruism. He forced his reason to accept what his intuition denied, viz., that all we have been, are or may become, are merely automatic results of physical feeding, breeding and battle. No wonder this great scientist declared Life an unsolvable riddle, intelligence a delusion, love essentially lust, and morality without the sanction of Nature. No wonder he declared, "I wash my hands of Nature."

Upon just such extraordinary hypotheses as these, physical science attempts to account for and to explain Man. These are the theories which, no less than theological dogma, stultify intelligence, outrage conscience, and violate universal experience. The best intelligence of today declares that such theories, such assumptions, and such dogmas explain neither the psychical facts of Nature nor all of the physical facts. As a result, honest, unbiased, inquiring minds everywhere are seeking more satisfactory interpretations of the higher phenomena. Such investigators are moving forward regardless alike of dogmatic theology and scientific skepticism.

Life

"What is Life? That is another question and the answer is one of the profound mysteries. Science has explored Life down to a single cell

of living matter, but what makes the cell alive is not known." — Book of Popular Science.

Disregarding all law and certain facts, Modern science attempts to create man, define life and death, and solve the riddle of the Universe. Modern science holds that man is the product of evolution, Life is the function of the body, and Death ends it all. Bichat said: "Life is the sum of the functions by which death is resisted." Richerand observed: "Life consists in the aggregate of those phenomena which manifest themselves in succession for a limited time in organized beings."

Herbert Spencer devoted the entire third part of his Elements of Psychology to a consideration of the question, and by slow and steady strides comes to the conclusion that: "The broadest and most complete definition of Life will be the continuous adjustment of internal relations to external relations." Then came the last word from William Osler, considered the greatest doctor America has produced. He wrote: "The studies of the physiologist and physiological chemist abundantly indicate that all vital activities are ultimately the expression of molecular rearrangements and combinations. Life is, therefore, the expression of a series of chemical changes" (Modern Medicine, P. 39, 1907).

Professor De Bois-Reymond, University of Berlin, was not satisfied with these scientific definitions of Life. He stated in his book, De Sieben Waltraelthsagen, that there are Seven Enigmas which modern science has not been able to solve, viz. 1. Origin of Life and its qualities 2. Origin of motion 3. Origin of consciousness and sensation 4. Origin of rational thought and speech 5. Origin of free-will 6. Nature of force and matter 7. Designed order of Nature. Materialist and evolutionist that he was, Reymond avowed that there are unbridgeable gaps in the theory of Evolution which destroy it. He was constrained to declare that Universal Evolution is not a science but a creed proclaimed by a "group of willful men." Recent discoveries of electrons and cosmic rays have increased the confusion of the physical sciences, and caused Ouspensky to write: "In reality, no one knows anything" (P. 408).

Science Changes With The Wind

That which is called Science today is a very different thing from what it was yesterday, or what it will be tomorrow. A few facts have been verified and recorded, and a few laws have been approximately formulated. But all this is subject to revision or even reversion tomorrow.

Give to so-called modern science the most liberal meaning claimed for it, and still it has no existence outside of man's imagination. In no sense does it stand for Cosmic Laws and Cosmic Processes. At best it is a reflection of these in the beclouded mind of man. Cosmic Processes deal with realities and are governed by unchanging laws. Man deals with phantoms and is ruled by fancies. True Science, in the basic sense, is knowledge of the manifestation of that Infinite Intelligence and Invisible Power which activate the Universe and regulate its course.

Atheists And Evolutionists

The world meets another stumbling-block in the fact that modern science is ruled by atheists and evolutionists who declare that all is physical matter and mechanical energy, and that the Universe is composed of blind and unknown forces devoid of an intelligent principle. Carrel stated that, in various ways, modern science knows there is a spiritual world, but is forbidden to enter it under that definition (P. 41).

Truth lies beyond the reach of those who, thus, are limited in their search and in the description of what they discover.

The Physical

Scientists are hidebound in the orthodox. The one point above all others on which they are most narrow-minded, is on the insistence that a physical theory of the Universe must be developed by logical reasoning. But logical reasoning, mathematics and similar methods will not explain what the mind of man cannot grasp. Unknown to man are the laws of the Invisible World. If modern science must develop a physical theory only, it will never progress farther than it has to date.

There are only four known major things that compose the Universe, viz., space, matter, power and pattern. Oxygen is said to be an electrical form of force. Oxygen is a gas, and gases may be liquefied by pressure, proving that some power holds apart what constitutes them. They liquefy and solidify at low temperature, or when heat is extracted from them.

Cold is the absence of heat. Heat consumes space by expansion and allows for contraction when it is extracted from the molecules of substance. In cases where a heavy gas is extracted from a lighter one, as in the case of water between 0 and 4 degrees centigrade, the lighter gas will expand from it, but will contract thereafter. No true solid is devoid of crystalline structure. The crystalline character of solids seems to show that lines of force exist which arrange the corpuscles so they appear that way.

The mathematical precision of formulas of chemistry which have to straddle and vary to fit into the atomic theory, are more easily accounted for when we realize that the inherent property of matter and force and their affinities and peculiarities, can be accounted for and are proportional to each other at all times for like conditions and amounts.

The Cosmic Circle

We have explained more fully in our great course titled IMMORTALISM that all elements travel in circuits. Electricity travels in circuits. In our own body we have particles which at some time were in those of our most illustrious antecedents from centuries past, and even some of the stars and planets. All forces and substances have their own circuit, which holds them in their orderly course and shape. When the required amount of force is applied, an object may be moved from its course and even disintegrated. Under the Law of Polarity, the Process of Transformation (Creation) operates in a Great Oscillating Circuit, from God, the positive pole, to Man, the negative pole, and back from Man to God to complete the circuit. The stream is continuous, flowing both ways, as symbolized by Jacob's Ladder, from God to Man and back again (Gen. 28:12). The transformation from Birth to Death has its counterpart in the change from Death to Birth. As Life manifests on the physical plane, the change is termed birth; and as it passes on to the spiritual realm, the change is termed death. That change is not feared by those who understand (1 Cor. 15:15). If the truth were not hidden from the masses, it would be the end of modern theology.

Things spiritual we see and understand without Mind, not with our physical eye. Confusion rises from the fact that man is taught to expect material evidence where matter, as such, does not exist.

Can the eye of the dead see? It is Spiritual Man (Mind) that sees, not the physical eye. The eye is the instrument used by Spiritual Man in the function of vision. There is change, but no death. The theory of death is based on appearance, illusion, and was invented to give depots power over the masses. The change from being (invisible) to becoming visible is termed birth, and that from becoming (visible) to being (invisible) is termed death. In reality, man is never born and never dies.

In the Cosmic Circle of Life the "Principle of continuity extends from the physical world to the spiritual," wrote Drummond — and for that he became famous. He shocked the darkened world by voicing a factual truth taught by the Masters a hundred thousand years ago, and suppressed by Constantine in the 4th century A.D.

As bad climate, bad environment and bad habits devitalize the body and decrease its vibrations until it is unable to function efficiently on the life level, its polarity weakens, cosmic radiation is disrupted, and the body slowly sinks from the Great Life Circuit. Then appears the state of decrepitude termed Old Age, followed in time by somatic decay, termed death, and, under the law governing disintegrated matter, the body substance dissolves and returns to the Cosmic Reservoir (Eccl. 12:7). In its animated state, the body is subject to the Law of Life. It vibrates on the Life level under the power of the Animating Force of the Cosmos, with constant changes occurring in its substance under the double law of synthesis and analysis. The body is composed of universal elements which are constantly changing. They enter into and pass through the body, abiding in it for a time without losing their identity with the Universe to which they belong and of which they are part.

The Physical Plane

The Ancient Masters asserted that the four cosmic elements of air, fire, water and earth are the builders, sustainers and regenerators of the body. These appear as four phases of the Cosmic Ray as follows: 1. Thermic 2. Atmospheric 3. Fluvistic 4. Terrestrial

These Four Rays are charged with all the elements of the living organism, and many more. The four elements of oxygen, hydrogen, nitrogen and carbon constitute 97.10 percent of the elements of the organism. Four is the symbol of New Jerusalem. The Pythagoreans called Four "the great miracle; a God after another manner, a manifold; the Foundation of Nature." The fourth letter is Daleth (D); and Daleth is called the womb," for it is the feminine First Cause, as Aleph or No. 1 is the masculine First Cause. Four is the number of the physical plane; of stability; of the foundation stone upon which is reared Man's spiritual structure. It comprehends all powers, both productive and produced numbers. Two multiplied into itself produces four; and retorted into itself makes the first cube. The cube is the fertile number, the ground of multitude and variety. Thus, from one fountain flow the two Principle of temporal things, the Pyramid and Cube, Form and Matter.

In the ancient Egyptian Mysteries man's symbol was the Cube unfolding, and Six becoming Seven, or the Three crossways (female) and Four vertical (male). The Mysteries used the interlaced triangles, of the Six Pointed Star, as a symbol of generation, the union of fire and water, the male and female. The four phases or planes extending from God to Man were termed the Etheric Body by the Egyptian Masters.

Flowing from the Godhead, each phase of the Etheric Body serves as a vehicle for the one next above, which order enables the Spirit to function on all planes. As the phases of the Etheric Body reach the limit of descent, as in the case of the dense physical body, it is via the last phase in the line. Nothing can reach the various planes, from the highest to the lowest, except from the plane next above. As the Etheric, in its four phases, is the one next above, all changes of the chemical body take place through the Etheric. Man's physical or chemical body and his vitality are not derived from what he eats, but from the Etheric Body that flows from the Godhead, or Spiritual Sun, as the Ancient Masters called it. They did not worship the Sun as God, but as God's primary representative. Man's chemical body consists of three subdivisions, viz., solids, liquids, and gases.

Four Cosmic Bodies

The next higher and less dense body, as we ascend toward the Godhead, are the etheric planes, divided as to density into four phases as stated, each having a different function, and all reflected to the chemical body of man. Counting the phases from above downward as No. 1, and the lowermost next to the dense chemical body as No. 4, this fourth phase of the Etheric is known as the Chemical Ether. Through it operates the force that causes the animation and function of the dense chemical body. The third phase is the Life Ether. It controls the maintenance of the dense chemical body, and also the function of generation or propagation.

The second phase from above is the Light Ether. It elaborates the blood in man and beast, and the fluids and coloring matter of vegetables. The first phase from above is the Reflection Ether. It is the medium through which Thought makes its impression on the brain. It reflects the Mind and records the Thought Pictures permanently from its corresponding subplane of the higher realms. Without the Reflection Ether, Mind and Memory would be absent.

These four Etheric Planes, referred to often in the biblical allegories, constitute the Etheric or Infinite Body, from which vitality and substance flow to the dense chemical body of man. Ancient Science termed these four planes the Guardian Angels. They were symbolized as the Cherubim (Gen. 3:24). They are the four beasts of Daniel, and referred to in the Ezekiel as the Man, the Bull, the Lion, and the Eagle. Occult science lists them as the chemical, the etheric, the astral, and the mental planes. The Creative Circle appears in a drop of water, and the transformative processes that make cosmic substance visible in the form of water, ice, snow, steam, stone, are the same that make Man.

Vibration alone determines the visibility and invisibility of substance. The radio intensifies the rate of vibration until we can contact radiations that are otherwise beyond the reach of our sense of hearing. As we vibrate so we live. Our rate of vibration determines the scope of our consciousness. As we become able to increase our rate of vibration, we extend our scope of consciousness. How that is accomplished we explain in the course IMMORTALISM.

It is a limitation of consciousness in man that creates illusions which are not recognized by him as such. Otherwise he would know that Man is the image of God in Spirit, and that he exists in infinite space and time. Man was never created as that term is understood and as theology teaches. Man always was, and he always will be. Man is that part of Creation in which appears a complete microcosm. Man is a microcosm of the Macrocosm. All is in man as occultism asserts. He is formed on that pattern which exists in the Spiritual Realm, and that pattern is the Godhead. Man is the physical instrument through which God is apparent, manifest, and revealed on the visible plane.

Rise Above The Material

We rise above the material level in our mind. As we learn more of the real Life, we cease to identify ourself with our physical body and our material surroundings, and direct our thoughts to the Eternal which lies beyond the temporal. When the ethics of modern science require that everything must be considered and explained by a physical theory as physical scientists feel hidebound to do, instead of recognizing and considering the higher forces of Life, psychical and spiritual, we would be better off to remain ignorant.

If we reject the Real, confine ourselves to the realm of illusion, and limit our vision and knowledge to the physical aspect of being, as Modern science does, and judge all things from the basis of material experience, then we can never reach a clear conception of the higher realms of Life. For the result of searching for a physical theory will produce nothing but chaos and confusion, such as that which prevails at present in Modern science. It may be correctly considered as a Resurrection of sleeping man when the Light of Knowledge based on Truth penetrates his foggy Mind and thus raises him up from the Grave of Materiality and Ignorance.

Physics And Chemistry

Ancient Arcane Science held that man contains within himself all the universes, systems, planets and globes. He is the microcosm of the macrocosm, partaking of all life. He is constituted to correspond with the visible and invisible world, the spiritual and the terrestrial, the eternal and the temporal. This man, this superior living being, medical art and Modern science regard as a purely physical entity, and have twisted practically every fact of existence in the attempt to reduce man to the low level of physics and chemistry. Modern science stubbornly refuses to accept the doctrine of a Principle of Life and of Intelligence.

Carrel discovered enough in his work with the living organism to realize that man is much more than a physical entity. He wrote: "The illusions of the mechanicists (and physicians) of the 19th century, the dogmas of Jacques Loeb, and the childish physico-chemical conceptions of human beings, in which . . . physiologists and physicians still believe, have to be definitely abandoned" (P. 108).

That would mean the abandonment of all medical theories and writings as to man and of the treatment of his maladies.

Intelligence

Carrel worked hard and twisted many statements in his attempt to support the theory of science that *"the world is composed of blind and unknown forces."* He made that statement on page 16 of his Man The Unknown, and then on page 121 he said, *"the existence of Intelligence is a primary datum of observation."* These statements definitely conflict; and when two statements do not agree, it is certain that one is erroneous. Carrel made the first one to support the theory of "medical science," while the next is an admission of fact that was forced upon him by his own positive findings. His conflicting statements concerning the basic principles of Life and Living lessen the value of his work and reveal the distorted condition of his mind, which is the common result of medical training. Sound and certain conclusions are reached only by rigid consistency of thought, which demands that we proceed in our processes in a direct and undeviating course through infinite time to infinite results. If there is definite evidence of the existence of a Principle of Cosmic Intelligence so patent as to be a primary datum of observation, then the world is not composed of blind and unknown forces.

In regard to the Cosmic Principle of Intelligence existing, Carrel said: "The body perceives the remote as well as the near, the future as well as the present (P. 197). An organ is engendered by cells which, to all appearances, have a knowledge of the future edifice (structure) (P. 108). The innate knowledge of the part they (the cells) must play in the whole (body) is a mode of being of all the elements of the body (P. 107). The cells (of the body) seem to remember their original unity, even when they have become the elements of an innumerable multitude. They know spontaneously the functions attributed to them in the organized whole (P. 106). Individual cells appear to act in the interest of the whole, just as bees work for the good of the hive. They seem to know the future; and they prepare for this future by anticipated changes of their structure and functions (P. 226). Cells (of the body) are like bees erecting their geometrical alveoli, synthesizing honey, feeding their embryos, as though each one of them understood mathematics, chemistry and biology, and unselfishly acted for the interest of the entire community" (P. 107).

A world in which its parts so definitely exhibit their patent and uncanny powers is far from being "composed of blind and unknown forces." When Carrel made that statement, he meant that the world is devoid of Intelligence. Then he proceeds to show that his statement was erroneous by declaring that a definite Intelligence appears in the operation of every part of the body, down to the invisible cell.

A force is neither blind nor unknown when its work is so obvious and orderly as to be a datum of observation. In that case the mysterious, invisible force may be clearly seen in the mind, being understood by what it produces and the kind of work it does (Rom. 1:20). To hold that such Force is blind and unknown is self-imposed ignorance. In this case, it is a position taken in order to avoid the necessity of acknowledging the existence of a Supreme Principle of Existence.

Professor W. F. G. Swann, director of the Bartol Research Foundation, recognized the existence of Cosmic Law and Order, and said, "Viewing the Universe as a whole, I cannot escape the fact that it is of intelligent design."

It seems inconceivable that scientists deny the existence of Supreme Intelligence and then declare that "the Universe is orderly, understandable, and of intelligent design."

Fourth Dimension

The Spiritual Power of the Cosmos seems to be Omniscient. As one scientist expressed it: Every part of the Universe acts as if aware of what is going on in every other part. Einstein's theories, now generally accepted in principle, indicate that the Cosmos does not op[erate] according to purely mechanical laws. It appears to act according t[o the] power of One Omniscient Mind. The mechanical, materialistic theo[ry of] science as to the Universe, depicted it in three dimensions — le[ngth,] width and height. Our minds and words are unable to picture fo[ur or] more dimensions.

The law of relativity and kindred theories now show tha[t the] Cosmos cannot be limited to three dimensions. Four dimen[sions,] sometimes seven, are required in the mathematical calculatio[n of] scientists. This kind of mathematics was formerly considere[d an] abstraction — a plaything for mathematicians, having no applicati[on to] the Universe. Now it is found to be the Law of the Governing Mi[nd by] which the Universe and all its parts are directed.

We learn that space and time are not what they seem[. Our] materialistic conceptions are erroneous. Words have not yet [been] invented that can describe the actual facts of existence. That is[, the] recent discoveries can be understood only by advanced mathemat[ics] and cannot be explained to others. The average layman can know [only] that time and space, man and the Cosmos, as we know them by ou[r] physical senses, are illusions.

One noted astronomer wrote that all objects "exist in the M[ind of] an Eternal Being." Then we are simply atoms in the Mind of that [Being] and, like the atom, simply a center, somewhat more elaborate, o[f] thoughts. Scientific discoveries are bringing us back to the doctr[ine of] the Ancient Masters, that "in the beginning was the Word, and the [Word] was with God, and the Word was God. . . . And the Word was [made] flesh" and the flesh was man, made in the spiritual image of God a[s] described in the ancient scriptures. That is all there is. That is a[ll that] remains of man and every material object when traced by scie[nce to] ultimates. As we trace material substance to its origin, it fade[s and] vanishes into the Great Invisible World.

Modern scientists and evolutionists and medical docto[rs are] definitely atheistic. They had rather grope in darkness and self-im[posed] ignorance than acknowledge their error by recognizing the existen[ce of a] Supreme Principle of Life and Intelligence. Having denied the exi[stence] of God and declared that the Spiritual World of the Ancient Mas[ters is] gross superstition, they have no alternative but to assume that the [Cosmos] is composed of blind and unknown forces.

So-Called Law Of Gravitation

Opinion and presumption are not science. At best they are only theory. Yet men of relatively little knowledge accept the theories, opinions and presumptions of greater men as though they were established laws, when they have already been discarded by the scientists who first suggested them. A typical case of this kind appears in the so-called Law of Gravitation. Oil poured into water floats on the surface; poured into alcohol, it sinks. In a mixture of the right proportions of water and alcohol, the oil remains in the middle of the liquid and assumes the form of a perfect sphere, the size of an apple or so, if the quantity of oil is sufficient. Such a simple experiment shows that the Newtonian theory of gravitation does not apply in the least, but that the principle of buoyance must predominate, as we can also see if we study colloidal chemistry and emulsions. Smoke is a colloidal suspension, only possible by a buoyant force. Modern science considers the so-called law of gravitation as an established fact, and thus applies the law to its work, as a result of which it encounters many difficulties and much confusion.

The so-called law of gravitation was stated by Newton in his book published in 1687. The real formulation of the law is as follows: "There are observed phenomena between two bodies in space that can be described as presuming that two bodies attract each other with a force directly proportional to the product of their masses and inversely proportional to the square of the distance separating them."

The formation of the law as recognized by Modern science is as follows: "Two bodies attract each other with a force directly proportional to the product of their masses and inversely proportional to the square of the distances separating them."

In the second formation, as recognized by modern science, the fact is conveniently and entirely forgotten that the force of attraction is merely a fictitious quantity accepted only for a description of phenomena.

In his text-book of Physics, Professor Chwolson wrote: "The tremendous development of celestial mechanics, based entirely on the law of universal gravitation taken as a fact, made scientists forget the purely descriptive character of this law, and see in it the final formulation of an actually existent physical phenomenon" (Vol. I, P. 182).

And thus it goes with Science. Newton never established the matter as a fact that bodies are actually attracted by one another, nor did he establish the reason why they are attracted or through the mediation of what. One student asks, "How can the sun influence the motion of the earth through the void of space?" "How in general is it possible to conceive action through empty space?"

The so-called law of gravitation does not give an answer to these questions, and Newton himself was well aware of that fact. He and his contemporaries, Huygens and Leibniz, gave warning against attempts to see in Newton's theory the solution of the problem of action through empty space, and regarded this theory merely as a formula for calculation.

In the first half of the 19th century the idea of action at a distance reigned supreme in science. Faraday was the first to point out the impossibility of the admission that a body should, without mediation, excite forces and produce motion to a point where that body is not situated. Leaving aside the question of universal gravitation, he turned special attention to the magnetic and electric phenomena and pointed out the supremely important part played in these phenomena by the intervening medium which fills the space between the bodies that appear to act upon one another without mediation.

We have discussed this matter somewhat at length to show the student that even the so-called law of gravitation, recognized by modern science as positive and definite, is nothing more than a presumption, a fictitious quantity, advanced by Newton only for a convenient description of phenomena.

End Of Creation Theory

Darwin's work put an end to the unsound theory of creation according to modern theology. But he merely reverted to the doctrine of ancient philosophy. The Ancient Masters held that "man is related to the animal." These Masters said that man sums up and crowns the series of beings, reveals all the divine thought in the harmony of his organism and the perfection of his form. For he is the living image of the Universal Soul or active Intelligence. Consolidating all the laws of evolution of the whole of Nature in his own body, man dominates and rises above physical nature, in order to enter freely and in full consciousness into the boundless kingdom of Spirit. Experimental psychology, which is grounded in physiology, has shown a tendency to become a science again, as in the days of the Ancient Masters. It has brought contemporary scientists to the threshold of another world, the Spiritual World, in which spiritual laws hold sway, but without the analogies ceasing to hold good.

We are hearing the mention of medical investigations and discoveries of animal magnetism, somnambulism, and all the various mental states of the subconscious self from lucid sleep through double vision on to a state of trance. So far modern science has been merely feeling its way into this mysterious domain, where Ancient Science went straight for the goal because it possessed the necessary principles and interpretations. Modern science is shocked by having recently discovered in this invisible realm a whole series of facts which appear marvelous, astonishing, and inexplicable. For these discoveries clearly contradict the materialistic theories under whose sway it acquired the habit of thinking and working. There is nothing more instructive than the indignant incredulity of certain physical scientists when brought face to face with all those phenomena that tend to prove the existence of an invisible, spiritual world. In this the reputation of these material scientists faces a terrible test. What is implied by the simplest phenomenon of mental suggestion at a distance and by pure thought, a phenomenon continually being demonstrated in the annals of magnetism, as a mode of motion both of mind and will, outside of physical laws or the visible world. Thus, the door of the spiritual world has been thrown open. In the higher phenomena of somnambulism, this world appears in its full extent.

The Secret Doctrine of the Ancient Masters was not merely a science, but a religio-philosophy. It was a science, philosophy and religion of which all the rest are nothing but preparations or degeneracies, partial or erroneous expressions, according as they proceed to them or turn aside from them. A theology imprisoned in dogma, and a science chained to materialism, cannot fill the desires of the human heart.

Nature

The Ancient Masters, scientists of the highest degree, saw that aspect of the Cosmos termed Nature in a much different light than we do, thanks to our type of education. We are taught that Nature does this and that, showing that the basic meaning of the word is disregarded, while the word itself is distorted in its use to serve the purpose of organized institutions that control the Nation.

Nature does nothing. Nature is the made, and not the maker. Nature is the Visible World, composed of cosmic substance and cosmic force, and under the direction of cosmic intelligence (law). The word Nature evolved from Nasci — to be born. Nature is born the same as man is, Nature is the existing order and products constituting the Material World, of which man is a part. The Ancient Masters considered Nature the work of the Great First Cause, the existence of which is admitted by the same physical scientists who deny in the next breath the existence of a Life Principle and a Principle of Intelligence.

Herbert Spencer f 1820-1903), strong exponent of the modern philosophy of Materialism and Evolutionism, wrote: "In our search for a cause, we discover no resting place until we arrive at the hypothesis of a First Cause; and we have no alternative but to regard this First Cause as Infinite and Absolute. These are inferences forced upon us by arguments from which there appears no escape" (First Principles).

Nature constitutes the Visible World. It is the Book we read as we strive to learn what we can of the mysteries of the Universe. That is the only book which God ever wrote. It is His Word, and contains the only Basic Truths in the Universe. More than 2,500 years ago Confucius said: "Does God speak? The solar system holds its orderly course, the earth fails not to revolve on its axis, the seasons pass in regular order, and all things continue to Jive and grow according to positive law; yet, I ask; Does God speak?" Man is part of Nature, and produced exactly as are all other parts of Nature. He is constituted of the same elements, and all phases of cosmic law that apply to nature, also apply to him with equal force and effect. Nature is that which is. True science must be Natural Science. For it is based on God's work, and is, therefore, God Science. It is the only Science that correctly records the true interpretations of what we read in the Book of God. Unprejudiced, logical reasoning is the consistent process of reading that Book.

In Nature, Natural Science, and Logic, we have subject, object and process, all indissolubly connected, illustrating the same principles and leading to the same conclusions from every angle of consideration. All must focus on the same point, the Point of Origin. Nature illustrates the Processes of Creation as well as the Products of Creation. The Products stand before us, and speak for themselves. The Processes are not so patent. They may be clearly seen in the Mind, being understood by the work they do and by what they produce (Born. 1:20). When these Processes are disclosed without bias or prejudice, then we are able to predict and/or determine the correct results.

How Cosmic Processes do their work, how they grow a plant, produce fruit, build men, produce good and bad health — that is the task to which the Ancient Masters directed their efforts, to learn how to apply certain conditions in order to produce certain results. Certainty is the product of fixed law. Anything done by cosmic processes is proof that it will always be done in the same way under the same conditions.

Variety of production comes secondarily from the same cause. Under the same conditions the same results are obtained. Under a change of conditions it is evident there must be a corresponding change of results. This is a certain fact whether in chemistry, mechanics or physiology. The causes of things are dependent in their action upon occasions or conditions as these bring into operation the Law of Production. That is the reason why cosmic processes can produce a plentitude of compounds from a paucity of elements. Whenever two or more elements come together under favorable conditions, the law of production being fixed, a compound is produced which is necessarily different from any one of the elements. But this compound is subject to change with every change of conditions, as the facts of chemistry constantly prove.

Two Kinds Of Facts

Exact science must present a true picture of the Processes as well as of the Products. That picture must be based upon actual facts and not fancies. Nature is constituted of facts, and is based on Principles. Nature is the product of Principles. Facts fall into two classes. In both Nature and Science there are (1) Facts Observed, and (2) Facts Inferred. Ancient Science established their existence and traced their connections. Facts observed are known as phenomena — the things we see and feel, the symptoms, as the doctor would say, constituting the Visible World.

Here at the starting point is where we encounter the first error of Inductive Science, and from there on physical science is hopelessly lost. While physical science denies it, all experience proves that the Visible World is not the Real World. We term it the Shadow World, the World of Illusion, in which the unenlightened man shares his knowledge with the beast, and upon which plane all men are in the same class until they see the light by finding the Kingdom of God within the organism it has made.

The theories of Modern science are based upon what it finds in the World of Illusion. It believes only what it sees, and thus thinks that Life is the expression of a series of chemical changes. The chemical changes observed are effects only, and are produced by an invisible cause. Every cause is a power or principle, constituted of Force, operating under the control and direction of Law. The scientist who trust to observation for his facts, is looking at the wrong side of Nature. He lives in the World of Illusion but knows it not.

The great truths which have shaken society to its center, have always appeared insignificant to the superficial observer; while to the discoverer, the true philosopher, the comprehensive thinker, the True Principle is a Pearl of priceless value. To him who has attained the true vantage-ground by discovering the Principle, everything is clear, full and obvious. Speculation then gives way to knowledge, and empiricism to the certainty of science. It is the Real World, the hidden causes of things, the invisible forces and the laws of their operation that we must discover if we should know the mysterious secret facts of Life and Death.

See The Invisible From Within

Facts inferred are considered as conclusion reached by a process of logical and consistent reasoning from Facts observed. The Ancient Masters explained it thus: "The invisible things of God from the creation of the world are clearly seen (in the mind), being understood by the things that are made" (visible). --- Rom. 1:20).

Man is possessed of Reason. The first task of Reason is to infer the Invisible from the Visible, the Spiritual from the Physical, the Unknown from the Known. In this respect, Reason is a faculty never used by modern science. Inferences are special conclusions from particular circumstances. They serve as links in the chain of reasoning. Had they been considered by modern science, it had never developed nor accepted the theory of Evolution. Observation shows that phenomena are effects. Reason declared that effects are necessarily the product of Causes. Hence, the Facts Inferred become at once the great necessity of an exact Science. Modern science positively disregards Facts Inferred. It holds that they are of no value until definitely verified. But it is this verification that makes them to be Facts of Inference.

Facts Inferred would seem to be the circumstantial evidence of Science, and hence the most reliable form of scientific knowledge. But as they are barred from consideration by the ethics of Modern science, it can be understood why the Life Principle, Cosmic Intelligence and the God of Creation are rejected by Modern science as rank superstition.

Science Of Man

Nature is composed of Causes and Effects, constituting the Facts Inferred from the Facts Observed. Ancient Science traced by logical and consistent processes the exact relationship between these two classes of Facts, showing what Causes produce what Effects, in order that they might obviate, produce, control or at least explain, the Effects. Science must be practical to be useful and valuable. Its province is to confer power by developing Knowledge based on Truths and the Facts of Nature. When these Truths and Facts are disregarded, distorted, or suppressed, there can be no science. On this point Carrel wrote: "Evident facts having an unorthodox appearance are suppressed. By reason of these difficulties (suppressions), the inventory of the things which could lead us to a better understanding of the human being, has been left incomplete" (*Man The Unknown*, P. 40).

When the facts of Nature weaken or destroy the theory of materialism and evolutionism, they must be suppressed. It is the work of scientists in protecting their theories by the deliberate suppression of "evident facts having an unorthodox appearance" that prevents the Modern world from having A Science of Man.

Ancient Wisdom Destroyed

The Ancient Masters had a perfect system of Cosmic Science, developed over a million years ago, and handed down from one age to another. Various despots at different times tried to destroy this ancient system, and the work of destruction was finally consummated under Constantine and his successors. David Livingston briefly tells the story in these words: "It is a fact of history that this champion of Christianity (Constantine), motivated by the vain ambition to possess the exclusive religious power in the world, sent emissaries to the East, for the sole purpose of destroying their sacred scriptures, that he might hold this power over all people — controlling not only their temporal but their spiritual destinies also. "He had gained a knowledge of the priceless value of these Eastern Records, having previously summoned to the Council (of Nicea) some 2,000 learned men out of all countries, who came with their choicest revelations; and it was from the cream of these sacred writings that the Constantine (Christian) Bible was composed. Then, after obtaining from them what was adjudged to be the best, these emissaries were sent to obliterate, at any cost, the priceless records of tens of thousands of years, such as the Revelations of the Avestas of the Persians, containing the record of the life and teachings of Zarathustra 7000 B.C., the most ancient of the Ascended Masters known to the outer world; and the Vedas of India, containing the wonders of the great Brahma.

"It was to effect this monstrous destruction (of these ancient scriptures) that Costullius, a monk, plotted the destruction of the world's largest library at Alexandria, first in 390 A.D., which was only partially successful, but was finally consummated in 640 A.D., at the instigation of three Christian monks" (God's Book Eskra, P. 727. notes 11, 12; "Magic Presence," P. 353). — Book of David, P. 140.

The great stroke of destruction was planned under Constantine at the First Council of Nicea in 325 A.D., and as the work of destruction progressed, all of Europe slowly sank into a state of dismal darkness that has extended unto this day, giving the Christian world that system of science which we have been discussing. No public report of the proceedings of that convention was ever made and issued; but a secret report was compiled and is concealed in the files of the Vatican at Rome. It is shown to all priests when they have reached a certain grade in their work. The pseudo science of today is approximately three centuries old, and is the work of materialists, evolutionists and atheists. And we speak of Russia as a "godless country." Now with materialism exploded by recent discoveries, and the Spiritual World of the Ancient Masters shown to be a fact, great confusion reigns in all branches of physical science.

Prof. Hotema   *Man's Higher Consciousness*   *Lesson No.30*

Ancient Science

The Ancient Masters took the position that man could not solve the mysteries of the vast Invisible World without precise knowledge of the physical. They postulated that the Spiritual or Invisible World was directly connected with the Physical World, and that its center was within the living organism and was manifest in the intellectual principle and the animating force. That center they regarded as the Soul or Spirit, which was to them the Key that unlocks the mysteries of being. By concentration of their will-power and the development of their latent faculties, they attained to the Divine Center which they called God, The Luke refers to this as follows: "The Kingdom of God is within you" (17:21)

The Masters were not mere visionaries, paltry dreamers, nor the slaves of ambition. The world has never known greater men. They shone as stars of the first magnitude in the world of Divine Souls. The pseudo scientists of this age represent a barren generation of greedy men, controlled in their work by organized institutions, devoid of ideals, light of faith, not inspired by any desire to improve the lot of mankind, believing in neither God nor the Soul, and scorning the suggestion of a Future Life. And we call Russia a "godless" country.

The Ancient Philosophy was generally divided into four categories: 1. Theogony, or the science of absolute principles, identical with thy science of numbers as applied to the Universe, or sacred mathematics 2. Cosmogony, or the realization of eternal principles, 3. Psychology, or the constitution of man, and 4. Physics, the science of kingdoms of terrestrial Nature and of their properties.

The essential principles of the Ancient Doctrines may be formulated as follows: Spirit is the only reality — a fact now proven by modern science but not admitted. Matter is condensed spiritual substance, and the theory of materialism is bankrupt. Creation is an eternal process, never beginning and never ending. It continues just as life continues. The microcosm (man) by reason of his constitution, is the image and mirror of the Macrocosm (Universe), divine, human, and natural world. It is for this reason that man, literally the image of God in spirit, can become his living world.

Spiritual Kingdom Within

Gnosis, or the rational mysticism of all times, is the art of finding God in oneself by the development of the occult depths and latent powers of spiritual and physical consciousness. To explain light, magnetism and electricity, it was necessary for modern science to posit the existence of a material substance that is subtle and imponderable, filling all space and penetrating all bodies — Matter which modern scientists call Ether. This is another step in the direction of the ancient philosophical doctrine of the Soul of the Cosmos, Universal Spirit.

Ancient Science was the logical development of discovered Truths and Facts in Nature, from which the Masters deduced a scientific system of philosophy and religion that harmonized and agreed with Cosmic Process and Laws. For the purpose of preservation, their system was concealed from the eyes of the world in a puzzling mass of symbols, allegories, parables, fables and fiction.

The written records of this system are termed the ancient scriptures. They were not prepared to teach the masses, but to conceal and to protect the Ancient Wisdom from loss and destruction. Being unable, without the key, to decipher the symbols, allegories, parables, fables and fiction, the despots would regard the whole thing as of no harm nor value, and not molest it. Otherwise the Ancient Wisdom had been utterly lost.

Man is the subject of the ancient scriptures, and the Science of Man is the Key to them.

The clergy knows nothing of the Ancient Wisdom concealed in the scriptures, or it would not teach that, "Except a man be born of water and of the Spirit, he cannot enter into the kingdom of God." (John 3:5). The Spirit in man is the Image of God; and in the body of every man is the kingdom of God, as stated in the Luke (17:21). That is the occult reason why man's body is more than form. Spiritual life means the development of spiritual consciousness within, and not the observance of some creed.

Our defective understanding, our distorted vision, our social bondage, our controlled minds, — these are the natural products of modern education and training, and are designed to crush the Spirit within and fit man in the social pattern created by organized institutions.

We rise to great heights when we know more about the kingdom of God within, and realize that our body is only a material instrument through which the Spirit functions on the material plane.

It is Cosmic Spirit in the kingdom of God (human body) that animates it, that produces physical consciousness in the brain, and enables man to see, hear, taste, smell, feel, think and reason — the Seven Mysteries of Life — that Power which has not yet been seen, nor heard, nor touched, nor named, and the very existence of which modern science denies. Self-imposed darkness.

Alipilli, of the mystic school of Alexandria, said: "I admonish thee that desirest to dive into the inmost parts of Nature, if what thou seekest thou findest not within thee, thou wilt never find it without thee. The universal orb of the world contains not so great mysteries and excellencies as a man, formed by God in His own spiritual image. And he who desires the primacy among the students of Nature, will nowhere find a greater or better field of study than himself. So, with a loud voice, I proclaim: O, Man, Know Thyself In thee is hidden the treasure of treasures." Solar Logos, P. 27.

The Ancient Voice

Now that modern schools and colleges are studying textbooks that teach the theory of materialism and evolutionism, with encyclopedias asserting that even the whole "world has come to accept evolution," the former leaders whose work did so much to produce this sad state, are finally renouncing the theory as false.

Professor J. S. Haldane, eminent English astronomer, wrote: "Materialism, once a plausible theory, is now a fatalistic creed of thousands; but materialism is nothing better than a superstition, on the same level as a belief in witches and devils. The materialistic theory is bankrupt." The theory of materialism has been utterly exploded by recent discoveries that Matter, as such, has no actual existence. Everything in the Universe is composed of Spirit in various forms and states of crystallization, exactly as the Ancient Masters declared. The doom of Materialism and Evolutionism has at last been sounded. Inductive science and physical science are left stranded on the barren rock of speculation.

They would never have been born had the founders of modern theology not destroyed the God Science and Philosophy of the Ancient Masters in order to enslave the mind of the masses. That wanton work of destruction of the sacred Ancient Wisdom was so well carried out, that Archbishop Chrysostom, in the middle of the 5th century A.D, boasted: "Every trace of the old philosophy and literature of the ancient world has vanished from the face of the earth" (Bible Myths, Doane, P. 436).

That boast was practically true for more than fourteen hundred long years of darkness. But nothing that has been can be completely destroyed. The light began to dawn again most unexpectedly in 1796, when Napoleon's army, digging trenches in Egypt, chanced to unearth the now famous Rosetta Stone from the soil of the Nile delta, where it had been deeply buried in the 4th century A.D. by Constantine's army of destruction. Then followed the marvelous work of Champollion in deciphering the stone's cryptic hieroglyphics. The world was amazed, for that decipherment released the Ancient Voice which Chrysostom had boasted and believed was forever silenced. And such a startling story as it told. It showed that the Masters of antiquity were not superstitious heathens, but scientists of the highest degree.

Now that Ancient Voice, which despots tried so hard to silence, speaks again from the dim and distant past, with all its mighty power and proficiency, in our MASTER COURSE titled...

The Glorious Resurrection

The student will find in that course the most complete compilation that has ever been produced of the Cosmic Science, Philosophy and Religion of the Great Masters of the ancient world.

Lesson No. 31
The Sacred Science

Man's Higher Consciousness proper ends here. The following remarks are appended to show more clearly how the masses are deceived by established institutions which suck the blood of the toiler who sweats to live, and trusts his teachers because he is led to believe that they are honest and represent worthy organizations. In the Sacred Ancient Mysteries, suppressed by Constantine and his successors, the Gnosis, or Ancient Arcane Science, was guarded with jealous care, and imparted only to a comparative few who were found worthy of initiation, according to the maxim, "Many are called, but few are chosen."

The Apocalypse (Revelation) is the last book of the New Testament. For several centuries an attempt was made to give it an historical interpretation. Failing in that, it was next interpreted as prophecy of the future. As it appears in the Bible, the Apocalypse is shot through and through with spurious interpolations.

Dangerous Invention

The invention of printing put a stop to biblical forgeries, and caused Cardinal Wolsey of London, in a convocation of his clergy in 1474, to say: *"If we do not destroy this dangerous invention, it will one day destroy us* (Doane, Biblical Myths, P. 438).

After the art of printing was invented, established institutions have sought to control publishing. This was done at first by licensing the printers. As more men learned the art, clandestine shops printed books that contained much revealing truth. This "danger" was met by the "List of Forbidden Books." Such books were burned when found, and in many instances the author was burned too. The gullible public think this could not happen now as all people want knowledge based on truth. The facts show otherwise. In the New York Herald of May 7th, 1901, the Rev. Harney, conducting a mission in St. Peter's Roman Catholic Church, New Brunswick, New Jersey, is quoted as stating: *"I do not doubt, if they were strong enough, that the Catholic people would hinder, even by death if necessary, the spread of heretical errors among the people, and I say rightly so."*

In this case, the "spread of heretical errors" means the dissemination of knowledge that feeds the Mind and frees the Man from the snares of the social pattern. We are attacking no creed, but are facing cold facts. If your creed is so contrary to the laws of the Cosmos that it cannot face facts, your faith rests on a mighty frail foundation.

More Pious Fraud

The Apocalypse is purported to be the work of "St. John The Divine," who is also accredited with the fourth gospel. The Bible says that he was an ignorant, unlearned fisherman (Mark 1:19: Acts 4:13).

These two books are not the work of one man. The fourth gospel was compiled from a biography of Apollonius written by Damis, a Greek historian and the most beloved disciple of the Master he loved so well. The compiler of the New Testament deleted the name of Apollonius and inserted the name Jesus. Then the original manuscript was to be destroyed—but it was not. Damis was a prominent writer, and in the fourth gospel he defined Life (6:63) in the words of his Master. That is more than modern science has been able to do.

The Apocalypse contains in symbol and allegory the Science of Man and the Secret of Regeneration. Modern science has not yet gone that far. The Apocalypse is the greatest work in symbol and allegory ever written. It deals with the fable of Genesis as to the Fall of Man. Both books refer to nothing but the human body and its deepest functions, about which modern scientists admit they know almost nothing.

Spiritual Life

Modern science is based on the theory that the atom was solid and eternal, indivisible and indestructible. It has been discovered that the atom is as empty as the air appears to be. If an atom were magnified to the size of a large room, internal whirling particles could be seen. According to science, these particles are whirling centers of energy.

The whirling particles in the atom are intangible. They are called electrical. That means little, for no one knows what electricity is. The word spiritual could as well be used, perhaps better. Material science boycotts the terms spirit and spiritual because they were used by the

Ancient Masters to designate Invisible Cosmic Force. Things spiritual modern science condemns as heathenish superstition.

It seems more sensible and scientific to speak of Spiritual Life within us than to term it Electrical Life. Electricity appears as undirected force without intelligence that makes nothing. Spirit appears as directed Cosmic Force with Intelligence that makes everything. Proof of that appears in things made. If not so made, then how made? To that question science is silent.

Atoms are kept in motion by a force termed energy by science. The Ancient Masters called it Spiritual Intelligence, and declared that it is omnipresent pervading all things. Then came the curtain of darkness in the 4th century A.D., and men were led astray into believing that God was far away, beyond, separate. Modern science was fated to prove the omnipresent existence of God while denying Him.

Regeneration

The Apocalypse is the despair of theology. The "infallible" church does not pretend to understand it. The clergy admit that it must be regarded as an unsolved, and possible insoluble, enigma. They translate its title "Revelation" - yet to them it reveals nothing. Man is the subject of the ancient scriptures, and the Apocalypse is the Key to the secret of Regeneration.

"O, Man Know Thyself In thee is hidden the treasures of the Universe." — Alipilli.

Ye that have followed me in the Regeneration (Mat. 19:28) is a loose statement that leaves the student in darkness. No evidence appears to show when, where and how any "regeneration" occurred in those to whom that remark was made. Incomprehensible as the Apocalypse appears to the exoteric student, however great his intellectual attainments, keen his mental acumen, and vast his store of erudition, to the mere typro in the Ancient Arcane Science the general tone of the Apocalypse is clear and sweet. It is not clear to the exoteric student because it deals, in heavily veiled symbolical language, with the Human Body, its Seven Principle Spiritual Centers and its Twelve Major Functions.

The attainment of spiritual knowledge is in effect the process of reviving the memory of the intimating Ego in relation to the supernal world before it became immured in matter. The memory of things divine can be resurrected only through the action of the Parakletos, the Regenerative Power. Hence, this aspect of the Nous (Mind, Intelligence, Consciousness), is said to hold in its grasp the Seven Stars and to move among the Seven lamp-stands.

According to the Ancient Masters, all true knowledge is derived from the "recollection of the things in which God abides."As the sun enters each sign of the zodiac, it is said, astrologically, to conquer the sign and to assimilate its particular quality. The same is said of him who raises up the Kundalini, the Coiled Serpent, the Speirema, the Fiery Force, said to dwell in the Sacral Plexus. As it flows up through the Silver (Spinal) Cord, it activates the Seven Etheric Centers of the organism, and enters the Golden Bowl (Skull) (Eccl. 12:6). He who does that is the hero of the Apocalypse, and is called the Conqueror (Revelations 21:7).

Man's Decreasing Powers

Man is complete. His constitution is such that he corresponds with the two worlds, the spiritual and the physical, the eternal and the temporal. The various parts and organs of his body correspond to and harmonize with every force and element of this Universe. The neophyte was taught that scientific knowledge in the Ancient Mysteries, and it is contained and concealed in the Apocalypse. That was the Ancient Arcane Science, and that science had to be suppressed to keep man in darkness. If man knows that he is eternal, he has no need for priests and saviors.

Due to degeneration, civilized man has lost the conscious capacity that connected him with the higher world, and is rapidly losing the power that informs him of the lower world. His five physical senses are constantly growing more defective. Insanity is increasing at an alarming rate, and its pace is hastened by sexual debauchery. The vital essence that produces New Life will preserve the Old Life if not otherwise consumed, and increase all the body's powers. Not only is the Life Essence the most precious fluid in the body, but the gonad glands that produce it are the most important glands in the body.

The dark den of "sin and shame" in which the Productive Function of man has been immured for ages by the priesthood, will disappear as we feed the Mind that frees the Man. Those who are properly taught that the Seed Glands are the most sacred glands of the body, become able to discuss the subject without squirming in their chair and showing red in their face.

Dr. Serge Voronoff, who recently died at the age of 85, became world famous for his experiments in transplanting monkey glands into human beings for rejuvenation purposes. He thought he had discovered the secret of Perpetual Youth. He considered what the loss of the precious fluid of the genital glands did to eunuchs, castrated horses, hogs and other animals. He saw that the loss of the genitals and their fluid resulted in the emasculation of the animals. Whether man or beast, the result was always the same.

Voronoff reasoned that decrepit men are to all intents "eunuchs." They have dissipated their Life Essence, and the weakness and aging resulting have affected the whole body and brain. The genitals were no longer competent to elaborate the vital fluid the body and brain need, and the body sinks into decrepitude.

The Endocrine Glands are the Master Chemists of the organism. Upon their products depend the function of all other glands. The Gonad Glands are the leaders and controllers of the Endocrine System. They are the Life Glands and they produce the most refined and most vital fluid in the organism. Loss of their substance in sexual debauchery diminishes brain power and decreases man's conscious capacity. For that reason these glands are also called the "destructive glands." In the act of production, man sacrifices in no small degree his own vital force and substance. Far worse than this is his shameful dissipation of the Life Essence for pleasure only.

The Ancient Masters correctly termed this substance the Life Essence. They knew of the damage the body suffers because of its loss. Physiological facts prove they were right. Hence, we find that Chastity is written in words of fire on all the pages of ancient scriptures.

The Red Dragon

The profound allegory in the Apocalypse readily unfolds as we learn with what it deals. When we know it refers to man and his

Regeneration, we understand that it does not treat of historical events or of future mundane conditions. The woman clothed with the Sun, with the Moon under her feet, and upon her head a crown of Twelve Stars (Rev. 12:1), does not assimilate saviors or savor of vicarious atonements.

The Sun symbolizes the supreme power that stimulates the function of generation. The Moon symbolizes the Generative Life, fecundity. The Twelve Stars symbolize the twelve major functions of the body, typified by the twelve zodiacal signs, which the Great Mother has mastered through physical regeneration and spiritual power.

The dragon that stood before the woman, who was ready to deliver, to devour her child as soon as it was born (Rev. 12:4), symbolizes the destructive carnal lust in the blood of fallen man.

This constellatory symbol Is Dracon, the pole Dragon, which has seven distinguishing stars and which, as depicted in ancient star-maps, extends over seven of the zodiacal signs. In setting, it apparently sweeps a third of the starry sky down to the horizon. Microcosmically it symbolizes the Passional Nature of Man.

Two Laws Of Generation

It is known to few that Two Laws of Generation are definitely referred to in the ancient scriptures: (1) the Sprawl Law and (2) the Carnal law. Of these laws Paul (Pol, Apollo, Apollonius) said: *"I see another law in my (generative) members, warring against the (spiritual) law of my mind, and bringing me into captivity to the (carnal) law of sin (copulation) in my (generative) members."* — Rom, 7:23.

In the Sacred Ancient Mysteries, the great school of Arcane Science, the neophyte came face to face with the Fiery Dragon that bars the way to the higher Life. Its eternal message to man is: *"I am you Animal Nature; if you would pass on to the higher life, then me you must master"* (Gen. 2:17, 1 Cor. 7:1).

Generation And Death

Modern biology proves what the ancient scriptures assert; Generation is a tax on the organism, a sacrifice which in time destroys the organism that consumes its vital essence in the exercise of the foundation function. Biology shows that, as to various lower animals, it is the rule that the moment of generation is the moment of death. Some insects live long enough only to generate offspring, and then expire. With some, the act of generation is the act of expiration. Margaret Merely wrote "Oftentimes, in the lower though complex forms of life, the parent literally resolves its whole substance into reproductive material, the maturing of this material causing the death of the parent. For instance, in certain very low forms, the parent becomes a mere shell to hold the progeny, and when they mature, it bursts open to free them, and thus expires. In certain forms of eggs, the female disintegrates upon liberating the egg-cells; and even as high as the insects, the parent in some cases is sacrificed by the developing of offspring." Life and Love.

The butterfly flits about for a brief space in the morning sunshine, displaying its vestments of life, then mates, and dies. The male and female approach each other in rapid flight. They meet, embrace, and in a brief period their love-flight ends in death. In a few moments they exhaust their vitality and die. After several months of preparation for this fleeting expression of existence, the supreme sacrifice is made, and the creatures fall and expire. The male butterfly, "faint, sated with all that life can give, serving the purpose for which it is made, totally exhausted by this final supreme act," fades from the highest delight of its brief love-flight to an inexorable death. The female flies away, and after depositing her eggs on some plant, the foliage of which serves the purposes of her offspring which she is fated never to see, she also falls in death.

In the vegetal kingdom, there are many plants that die immediately upon completion of the means of generation. They grow, mature, bloom, produce seed, and die. Fruit trees that begin bearing young, soon exhaust themselves and fail to attain a normal growth of wood. Careful horticulturists prevent this by removing the flower-buds of their young trees. Some plants, as with the higher animals, have recurring seasons of generative activity, followed by rest. Even in such cases, early maturity means early death and prolific bearing means early exhaustion.

Production in both the vegetal and animal kingdoms represents sacrifice. This rule has no exception. It prevails throughout the entire living world. Sir E. Ray Lankaster in his "Longevity," emphasizes this principle. He found traces of it in animals of the highest order. The higher one goes, the more cases of it one finds. Breeders of birds know that birds used for breeding live not nearly so long as those not so used, all things being equal. James S. Gould, a high authority on cage-birds, wrote: "Canaries well treated will live from 15 to 20 years, and sing to the last. Those used for breeding seldom live longer than 10 years" (My Canary Book).

If growth is rapid and maturity soon reached the generative reaction is exercised early, followed by early decay and early death. If growth is slow and maturity comes late, the generative function is exercised late, and slow decay and long life result. Methuselah lived 187 years before he begat Lamech, and died at the age of 969. Nahor lived 29 years and begat Terah, and died at the early age of 148 (Gen. 5:25-7; 11:24-5).

The Law of Generation and Law of Longevity are reciprocal and compensatory in action, proving that the Ancient Masters knew their physiology when they wrote that man enters the shadows of death in the day that he consumes his Life Essence in the generative act (Gen. 2:17).

Seven Spiritual Centers

The Book with Seven Seals, 5th chapter of Revolution, is another allegory that deals with the body and its regeneration. This allegory represents the fact that Cosmic Processes work through Sevens in the development and arrangement of physical forms. The Ancient Masters considered the body, as well as the Cosmos, as geometrical figures, and that Sound and Number rule Cosmic Processes.

The seven colors of the rainbow reveal the substance contained in clear light; and the Seven Seals symbolize the Principle Spiritual Centers of the body, through which the etheric face functions. Cosmic Radiation is picked up or received by the Brain, the chief of these centers, as a man-made radio picks up messages in the air, and by the Brain is distributed to the other centers according to wave length, color wave, and the requirements of the different parts of the body needing the different waves; then finally distributed over the body through the medium of the etheric (nerve) fluid that passes out, over and through the nerves. As we explained under "Four Cosmic Bodies," the chemical body is entirely dependent on the etheric. As all primal forces flows from above downward from the positive to the negative, if we place any obstruction in the circuit of a current, it will retard the flow. When the vibratory rate of the etheric current it retarded, it fails to deliver the proper amount of vitality to the chemical body.

Interference in the circuit of the Life Current is the reason why man's body declines, degenerate, loses its spiritual powers, grows decrepit, and finally sinks in death. We have explained in Man's Higher Consciousness how and why the life-channels, through which flows the Life Current become obstructed, clogged. That diminishes the amount of vitality, causing the function of the cells to fall below the life level of vibration. That is death, as explained under the Cosmic Circle. You can learn more of these amazing truths from our great works, the various writings of Professor Hilton Hotema, in which appears the most complete description of the Ancient Arcane Science ever contained in one work.

The ancient scriptures, from Genesis to Revelation and all others not contained in the Bible, are a compilation of allegories, parables, fables, and symbols that deal only with Man, his fall and his redemption. The gloom of the Dark Ages was caused and created by a suppression of the Ancient Arcane Science by Constantine and his successors, and the substitution of a new brand of theology that was invented to enthrone the dictators and enslave the masses.

The work was well done, and due to the math and power gained by the dictators as a result, many suffering souls will cry out for Spiritual Light, but few will find it.

CPSIA information can be obtained
at www.ICGtesting.com
Printed in the USA
BVHW051252130223
658403BV00003B/387

9 781639 233083